# UNDERSTANDING
# INFRASTRUCTURE

# BIOGRAPHIES

**George Rainer,** PE, AICP, a principal in the firm of Flack & Kurtz, consulting engineers, has been involved in the infrastructure design of Roosevelt Island and Battery Park City, New York, including utilities supply, solid waste management, water supply and sewer systems. He also prepared energy master plans for the Grand Valley Urbanization Project, Colorado, and for college campuses in Tennessee, Massachusetts and New York. He has lectured at several colleges and developed the course on infrastructure at Pratt Institute.

**Samuel I. Schwartz,** PE, is Chief Engineer and First Deputy Commissioner of the New York City Department of Transportation. He has been overseeing the Department's $8 billion, 10-year program to reconstruct New York City's streets and bridges. He chaired the Williamsburgh Bridge Technical Advisory Committee which developed a rehabilitation and construction program to save the bridge.

**Nicholas Bellizzi,** PE, a traffic engineer and transportation planner, is a principal in The Hudson Partnership. His assignments have included a comprehensive traffic/transportation study of the George Washington Bridge and its approach areas.

**Sarelle Weisberg,** AIA, is Program Manager for Public Works in the Department of General Services, New York City Bureau of Building Design and Construction. She is adjunct Professor in the Pratt Institute of Architecture Graduate Program in Regional Planning and Urban Design.

**Anthony Nuciforo** is a consultant in Communications Technology. He specializes in the design and implementation of large scale telecommunications cabling systems, and is developing a telecommunications infrastructure for a university complex in the Middle East.

ii

**Ekkehart Schwarz,** AIA, AICP, is president of Schwarz & Zambanini, architects, engineers and urban designers. A graduate of the Technische Universitat of Berlin and the University of California at Berkeley, has over 15 years' experience in the planning, design and redesign of streets infrastructure systems.

**Adrienne Bresnan,** AIA, has served for 18 years in the capital projects division of the New York City Department of Parks and Recreation. Since 1988 she has been program manager for landmarks in New York's Department of General Services.

**Joseph Bresnan,** AIA, served for 20 years in New York's Department of Parks and Recreation, where he created the preservation plans for Central Park and Prospect Park, Brooklyn. He was then appointed executive director of the city's Landmarks Preservation Commission, where he has helped strengthen its role as an important land use regulatory agency.

*Adrienne and Joseph Bresnan, who are former employees of the Department of Parks and Recreation, City of New York, wish to acknowledge the opportunity provided by the Department to develop concepts and proposals, leading to a wider appreciation of urban open space and the riches of the public park system.*

**Ray Gordon,** AIA, is a principal of the firm of Mark A. Kates Architects. Before that he was project manager of an environmental sciences firm where he specialized in a number of large scale waterfront development projects.

**Joseph DePlasco** is a writer and historian working the Chief Engineer's office, New York City Department of Transportation.

# UNDERSTANDING INFRASTRUCTURE

## A GUIDE FOR ARCHITECTS AND PLANNERS

**GEORGE RAINER**

A WILEY-INTERSCIENCE PUBLICATION

**JOHN WILEY & SONS**

**New York** / **Chichester** / **Brisbane** / **Toronto** / **Singapore**

*Library of Congress Cataloging-in-Publication Data:*

Rainer, George.
   Understanding infrastructure : a guide for architects and planners
/ George Rainer.
      p.   cm.
   "A Wiley-Interscience publication."
   Includes bibliographical references.
   1. Infrastructure (Economics)—United States—Handbooks, manuals,
etc.   I. Title.
HC110.C3R25   1989
363—dc20                                                89-36569
ISBN 0-471-50546-3                                          CIP

Printed in the United States of America

10  9  8  7  6  5  4  3  2  1

# CONTENTS

## 2   SEWERS AND STORM DRAINAGE     15

George Rainer

## 3   SOLID AND HAZARDOUS WASTES     31

George Rainer

## 6  STREETS 99

Ekkehart R. J. Schwarz

7 **BRIDGES**                                                    **127**

Samuel I. Schwartz
Joseph DePlasco

## 9 RAIL/TRANSIT AND AVIATION 187

Sarelle T. Weisberg

## 10 BUSES 223

Nicholas Bellizzi

# FOREWORD

What distinguishes modern society from the past is the higher standard of living available not only to the upper class, but to the average citizen. A closer look at what makes our society more liveable has to include the infrastructure that has been put in place in this country and in other Western countries since the 1800s.

Society's initial concerns deal with law, order, and public safety. Clean water and sanitary disposal of sewage are essential elements within our communities. These are followed by the conveniences of electric power, telephone communications and, ultimately, transportation networks.

The marvels of science and technology separate us from the past, but, with the complexities of convenience arise the problems of maintenance, expansion, and affordability.

Our cities have grown exponentially as a result of the population moving from the farms into the cities and then from the cities to the suburbs. The necessary expansion and replacement of the decaying infrastructure has become a national crisis. Recent Congressional hearings on the infrastructure have illuminated the enormous costs facing this country for replacement and improvement. Costs are projected to be in the trillions of dollars.

Architects and urban planners have a sketchy understanding of infrastructure technology. Clearly, the future will demand a greater grasp and awareness. George Rainer and his colleagues have compiled a comprehensive and readily understandable volume that will enable the architect and other design professionals to perceive and be cognizant of the multiple ramifications of infrastructure technology.

TED PAPPAS

*Fellow of the American Institute of Architects*
*President, Pappas Associates, Architects, Inc.*

# PREFACE

The purpose of this book is to introduce current or emerging practitioners in the fields of architecture, urban planning, and urban design to the basic necessities that make a city function. Although there is a general understanding of infrastructure needs among students in the design professions, this book is designed to sharpen and clarify this understanding and to provide facts on which to base important decisions in the development of urban spaces. A broader awareness of the deteriorated condition of the infrastructure in the United States is emerging. The rebuilding will require a sensitive knowledge of the issues—technical, environmental, legal, fiscal, and political—that are involved in their development. The new and renewed lifelines will require planning, superior design, and funding that will not only get them built but will ensure that they are maintained for the duration of their useful lives.

Each chapter deals with a specific issue; first the topic is described as to scope and given a frame of reference with which design professionals can identify. This is followed by descriptions of both standard and innovative solutions; environmental, legal, and economic considerations are discussed. Finally, as applicable, a bibliography and/or a directory of resources is appended. The references and readings will enable readers to delve more thoroughly into a specific topic of interest. Practitioners who wish to stay up to date will find these sources of value also, as legislation and regulations change so frequently.

The book is not intended to make readers experts in the areas covered, but to allow them to be better prepared to discuss relevant issues with consultants working on a given project. All the contributors are practitioners in their respective fields, so readers are assured that the material presented will be of considerable practical value.

GEORGE RAINER

*Irvington, New York*
*November 1989*

# ACKNOWLEDGMENTS

When Sarelle Weisberg took over the course on Urban Infrastructure that I had originated at Pratt Institute, it was agreed that the notes we had accumulated should be made available to young design professionals everywhere. Her encouragement and contraibution made the project a reality.

My thanks to Adrienne and Joseph Bresnan, Ray Gordon, Anthony Nuciforo, and Ekkehart Schwarz for graciously consenting to expand their notes into the chapters that represent their expertise.

When a critique of the first draft pointed out the omission of bridges and buses from the topics covered, Commissioner Sam Schwartz with Joe DePlasco and Nick Bellizzi jumped on board enthusiastically, and their quick response is gratefully acknowledged.

Many thanks to my colleagues Sheldon Steiner and Al Collado for helpful comments on the chapters in their fields.

Ekkehart Schwarz wishes to thank Lee Home and Frank Zambanini for their contribution to the chapter on Streets.

It is fortunate that Stephen Kliment of John Wiley & Sons agreed that a basic book on infrastructure was needed. Many thanks for making this book possible. To Jenet McIver and Mabel R. Vaughan, our thanks for looking after the details of putting this volume together so efficiently.

G. R.

## PHOTO CREDITS

Chapters 1 & 2 : *Marion Bernstein, NYC Department of Environmental Protection.*

Chapters 3, 4, 6, & 11 : *Renata Rainer.*

Chapter 8 : *Eugene Stamm.*

Chapters 5, 9, 10, & 7 (except Williamsburgh Bridge) : *Port Authority of New York and New Jersey, through the assistance of Ruth Singer; Williamsburgh Bridge : NYC Department of Transportation.*

# INTRODUCTION

The development of infrastructure—or public works, as these elements of regional services are often referred to—first became necessary in response to urban conditions that developed toward the end of the nineteenth century as a direct result of the industrial revolution. One of the aims of reform movements for better living conditions that started about 1890 was the improvement of public health conditions. Water supply and sewer systems were developed for that purpose, followed in due course by transportation systems, electric power, and telephone networks. The systems were built according to good engineering practice to last for 50 to 75 years. No systematic method for repair or replacement of these systems was put into place. At the present time the useful life of these systems in the older center cities has been exceeded and we are faced with replacing them at a time when they are being used more intensively than ever and when their replacement cost is at its peak.

Since almost no coordinated planning for this infrastructure was undertaken until well into the twentieth century, these systems were designed without the facility for expansion and without the flexibility to adapt to changing social, economic, or technological conditions. Health standards have become more restrictive with our greater awareness of possible sources of pollution; there has been an increase in the U.S. standard of living accompanied by greater expectations for services, and there has been a vast increase in mobility—all at a time of reduced federal spending on infrastructure.

A conscious selection was made in the choice of issues to be included. Physical infrastructure has been called the aggregate of all facilities that allow a society to function. This is a very broad mandate; therefore, the term "facilities" has been limited to the actual lifelines made of pipe, brick, and mortar without which no building or settlement would be useful. Not included are schools, hospitals, jails, and so on, which are simply buildings with special public purposes that are served by the physical infrastructure described here.

Infrastructure must also be responsive to social objectives; by its very nature physical infrastructure serves the social purposes of health, safety, economics, employment, and recreation. The changes in our social objectives that occur from place to place or over time find their obvious expression in the body politic, which allocates the funds to pay for the physical infrastructure. It is with a challenge to architects, planners, and urban designers to improve the nation's physical infrastructure that we present you with the necessary tools in the pages that follow.

GEORGE RAINER

# UNDERSTANDING INFRASTRUCTURE

Croton Gatehouse Construction

# 1

# WATER SUPPLY

**George Rainer**
Flack & Kurtz Consulting Engineers

## INTRODUCTION

The reason for dealing with the water supply issue as the first topic in this book is the well-known fact that no development of any kind can occur without water. Most of the important cities of the world have been located on an ocean, on a lake, or along a river because water was needed for their development and to help them in performing needed functions.

## USES OF WATER

The primary reason for locating early habitations along waterways was the importance of water transportation. Where natural water courses were inadequate, canals were built. Tankers and barges remain the most economical means of transporting heavy goods. Water power can produce electricity by driving turbine generators; in earlier times and in present-day underdeveloped economies, mills are driven by water power.

The major uses of water are in irrigation, municipal, and industrial applications. In the United States, 34.5% of all the water that is consumed is used for irrigation. Agricultural uses predominate; the western states have been chronically short of water. The federal government has built dams to contain floods and to even out the water supply, but shortages persist. Riparian (water) rights, which run in conjunction with the land, are highly prized possessions in the west. A minor portion of irrigation is for horticultural use; these are the uses that are curtailed first in case of water shortages.

Industrial uses comprise 54.5% of water consumption in the United States. This includes all industrial processes, as well as the cooling of machinery such as engines, compressors, and turbines. Water is clearly a utility that the economy of any town or city cannot be without.

Municipal uses comprise the remaining 11% of water consumption. This includes potable water supply for drinking and cooking, water for cleaning and sanitation, and of ever-increasing importance in densely populated areas, water for use in fire protection (Solley et al., 1988).

Finally, there are the many recreational uses that water makes possible, such as swimming, boating, and fishing. Fountains and waterfalls enhance the aesthetics of the urban scene. Indeed, there would be no civilization without water.

## SOURCES OF WATER

The basic hydrological water cycle in nature consists of evaporation, precipitation, and runoff. Water evaporated from lakes, rivers, ocean, and vegetation rises to the upper atmosphere, where it mixes with dust and gas in clouds. It returns to the earth in the form of rain or snow and then runs off to ponds and lakes or soaks into the ground. Figure 1-1 shows the basic water cycle.

At a certain distance below the contoured surface of the land is a boundary called the *groundwater table*. The water table becomes visible at rivers and lakes, and comes close to the surface in swamps. The earth below this level, consisting of clay, silt, sand, and gravel, is completely saturated with water known as *groundwater*. The porous area between the water table and the surface contains a mixture of air, liquid water, water vapor, and other gases known as *vadose water*. The difference between the two types of water is that groundwater will flow into a well under the force of gravity, whereas vadose water will not (Cohen et al., 1968).

A layer containing groundwater at atmospheric pressure is called an *unconfined aquifer*. Where the saturated level is overlain by silty or clayey layers of low permeability, the layer is called an *artesian* or *confined aquifer*. Artesian wells drilled into this type of aquifer frequently discharge water at pressures above atmospheric.

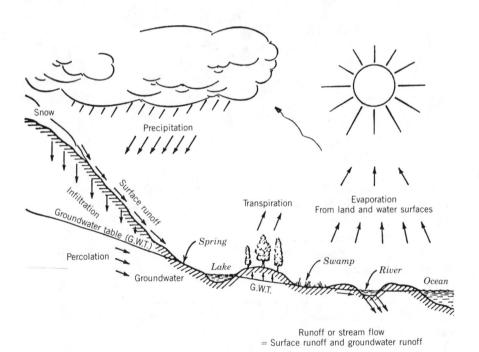

**Figure 1-1** The water cycle.

*Evapotranspiration* is the natural process by which water on or beneath the surface of the earth is returned to the atmosphere as water vapor. This process may be divided into three distinct elements: direct evaporation from wetted ground areas and vegetation, as well as from the zone of aeration shortly after the precipitation occurs; evaporation from permanent or semipermanent bodies of water; and the slow evapotranspiration of groundwater.

Maps showing the average annual rainfall for a given geographic area are available from the National Oceanic and Atmospheric Administration (NOAA). The fact that rainfall quantities vary from 3 inches per year in the southwest to 50 inches on Long Island, New York, to as much as 140 inches in areas of Washington State obviously affects the supply. Nationally, 78.3% of the water supply comes from surface sources: lakes, ponds, and reservoirs. The balance comes from groundwater sources described above (Solley et al., 1988). These underground reservoirs can be tapped by means of wells. When large quantities of water are withdrawn for industrial or process use, diffusion wells must be drilled to return the water to the ground to delay the depletion of the aquifer and to prevent the subsidence of the surface layers of the earth that could otherwise occur.

In large populated areas water supplies must be drawn from a wide area. The water supply for New York City comes from as far as 125 miles away; Los Angeles has to draw on sources 250 miles from its boundaries. As populations grow and as consumption per capita increases, demands become more intense. It has been estimated that by 1991 all U.S. cities of 50,000 population and above will require new sources of water due to population increase and/or depletion of resources (Congressional Budget Office, 1983). The challenge to plan for these needs is impressive.

## DISTRIBUTION OF WATER

The most basic method of distributing water is the bucket. In developing countries water is hauled out of wells in buckets and transported to the point of use. Bucket brigades are used in firefighting in areas that water mains do not reach. Aqueducts were first used extensively during the Roman Empire, as early as 100 B.C. Water from the Alps was fed by gravity to the population centers of the empire and large bridgelike structures were used to cross valleys or uneven territory. Some of these aqueducts are still in use.

Large quantities of water, such as those required for New York City, are distributed through tunnels. Approximately 1.3 billion gallons of water per day are delivered by gravity through tunnels up to 19.5 feet in diameter (McGough, 1983). Modern urban distribution networks consist of piping grids running in the bed of streets. Branch mains are usually fed from two feeder mains to provide service in the event that one feeder main is out of commission (due to a break, for example). Valves are provided to sectionalize branch mains to facilitate repairs or new connections. New York City water mains range from 4 to 12 inches in size, with some larger transfer mains. Potable water supply and fire protection service are generally served from the same mains. In commercial installations, however, the fire service frequently comes into the building separately; this allows the water department to shut off the potable water in case of nonpayment of bills, while also maintaining fire protection.

Water pressure in a system is determined by the elevation of the water source (e.g., lake, reservoir) and by the pressure drop in the distribution system. In New York City water pressure at the curb in front of a building ranges from 35 to 60 pounds per square inch, which is sufficient to deliver water to the top of a six-story building. Taller buildings must be provided with a pump to boost the pressure. In a typical installation the water is pumped up to a storage tank at the top of the building and then flows by gravity to the various users.

Where a gravity system is not adequate due to the relative elevations between the reservoir and the community, a water tower must be provided at the highest elevation. Water is pumped to the tower and then distributed by gravity from there. Some communities (e.g., San Francisco) provide a separate fire distribution loop at an elevated pressure, to facilitate firefighting. This water does not require the same purity as that in the potable loop, so it must be kept completely separate within the building.

Leakage can be a considerable problem in piping distribution systems, particularly when they have been in place for some time. Where all the supplies are metered, the final flow can be compared to the quantities put into the system. But where metering is not universal, estimating procedures must be utilized.

In northern climatic zones water supply piping must be protected against freezing. Prevailing ground temperatures can be estimated based on the coldest winter outdoor temperature. Burying pipes 4 feet below the surface will generally be adequate for freeze protection. Exposed pipes and tanks must be heat traced with electric elements.

## QUANTITIES CONSUMED

The consumption of water is influenced by a wide variety of factors:

*Climate.* People living in warmer climates take more showers, have more pools, and so on.

*Availability.* Where there is a shortage of water, there is a conservation ethic.

*Stage of development.* Industrialized countries have much greater requirements.

*Housing density.* Less open space requires less irrigation.

*Cultural preferences.* More or less personal hygiene is often a function of social mores.

As a broadbrush generalization, it may be stated that water consumption for cooking, sanitation, irrigation, and industrial uses ranges from 100 to 200 gallons per person per day. The New York City water supply system handles 1300 million gallons per day. With a population of 7.5 million people, this falls well within the normal range. Industrial uses have an inordinate influence on water consumption and industrial communities will always have a much greater need, on average, than that of a purely residential community (see Table 1-1).

Adequate water supply must be planned for—but nature does not operate according to human plans. In periods of limited rainfall, reservoirs may not be adequate to meet normal requirements. Contingency plans must then limit demand by law and must provide alternative or supplementary sources of supply. New York City, for example, maintains a pumping station on the Hudson River north of the metropolitan area which during periods of drought can supplement the water supply from the Croton reservoir system.

## WATER QUALITY

The quality of water is compromised by a variety of factors. Dust and spray particles are picked up by air movements over land and water surfaces and introduced into the atmosphere. Water vapor mixes with dust and gases in clouds and is then precipitated out of the atmosphere through rain or snow. The dissolved solids content of streamflow and lakes is also modified through physical, chemical, and biological processes. All of these cause the quality of surface water to vary throughout the year. Precipitation is also affected by pollutants in the air, generated as the products of combustion from fossil fuels, which then form acid rain.

Groundwater is in contact with rocks and dissolved minerals, which as a consequence appear in the water supply. The concentration of salts in water is an important characteristic. *Salinity* ranges from 10 parts per million (ppm) in rainwater, to 35,000 ppm in the ocean, to 250,000 ppm in the Dead Sea. Criteria for dissolved solids are as follows (Swenson and Baldwin, 1965):

| | |
|---|---|
| 1000–3000 ppm | Slightly saline |
| 3000–10,000 ppm | Moderately saline |
| 10,000–35,000 ppm | Very saline |
| More than 35,000 ppm | Brine |

**TABLE 1-1   ESTIMATED WATER REQUIREMENTS**

| Product | Unit Produced | Water Required | |
| --- | --- | --- | --- |
| | | Gallons/Unit | Gallons/Day |
| Buildings | | | |
| Office | Person | — | 27–45 |
| Hospital | Bed | — | 130–350 |
| Hotel | Guest room | — | 300–525 |
| Laundries | | | |
| Commercial | 1 lb work load | 5–8 | — |
| Institutional | 1 lb work load | 1–4 | — |
| Restaurant | Meal | 1–4 | — |
| Meat | | | |
| Packinghouse | 100 hogs killed | 550–600 | — |
| Slaughterhouse | 100 hogs killed | 550–600 | — |
| Stockyard | 1 acre | 160–200 | — |
| Poultry | 1 bird | — | 1 |
| Oil | | | |
| Oil refining | 100 bbl | 75,000–80,000 | — |
| Sugar | | | |
| Sugar refinery | 1 lb sugar | 1 | — |
| Paper | | | |
| Papermill | 1 ton | 40,000 | — |
| Paper pulp | | | |
| Ground wood | 1 ton (dry) | 5,000 | — |
| Soda | 1 ton (dry) | 85,000 | — |
| Sulfate | 1 ton (dry) | 65,000 | — |
| Sulfite | 1 ton (dry) | 60,000 | — |
| Textile | | | |
| Cotton bleacheries | 1 lb double boil | 25–40 | — |
| Cotton finishing | 1 yard | 10–15 | — |
| Silk hosiery dyeing | 1 lb | 3–5 | — |
| Knit goods bleaching with Solozone | 1 lb | 7–8 | — |
| Municipal requirements | Person | | 50–100 |

*Source: Water and Waste Treatment Data Book*, Permutit Co., Paramus, N.J., 1986.

In coastal areas, the pressure of seawater can create salinity in freshwater aquifers. This is caused by excessive pumping of the groundwater and can only be mitigated by the installation of diffusion wells, whereby water is forced into the ground under pressure.

*Hardness* is caused predominantly by the presence of magnesium and calcium (in the form of calcium carbonate) in water. Hardness keeps soap from foaming, leaves scale and residue on the surface of pipes and equipment, and is unacceptable in some industrial processes. Criteria are as follows:

| | |
|---|---|
| 0–60 ppm | Soft |
| 61–120 ppm | Moderately hard |
| 121–180 ppm | Hard |
| Over 180 ppm | Very hard |

*Color* and *turbidity* are related to the presence of suspended sediment in water. The U.S. Public Health Service (USPHS) sets an upper limit of 15 on a platinum–cobalt scale as being suitable for drinking, and any reading below 10 passes unnoticed. Filtration of drinking water usually removes turbidity, which is objectionable in concentrations above 5 ppm.

The balance between acids and alkalies, known as *pH,* is of importance. A pH of 7 indicates neutral water; above 7 the water is alkaline, below 7 it is acidic. This indicator is important for both industrial and potable supplies. A range of 6 to 8 is normal for fresh water; lower numbers cause a flat taste in drinking water.

The senses of taste and odor are closely related. Because both are subjective and personal, they lack scientific preciseness. However, disagreeable tastes and odors are commonly caused by living microscopic organisms, decaying organic matter, or industrial waste products such as ferrous oxides or sulfur which have seeped into the water supply. High mineral levels, on the other hand, add taste, odor, and color.

Temperature becomes a concern when industry requires water for cooling purposes and it is delivered at too high a temperature. Conversely, water returned to a river at too high a temperature can cause havoc with ecological cycles such as fish life. Also, hot water returned to an aquifer through a diffusion well can raise the water temperature over time to a point where its usefulness as a cooling medium may be lost.

Biological factors are significant in determining the quality of water. *Biochemical oxygen demand* (BOD) is a measure of the amount of oxygen required to remove waste organic matter from the water, and therefore provides an index of the degree of organic pollution. This measure is discussed in more detail in Chapter 2. Over the last 30 years a vast number of synthetic organic compounds have been developed and placed into service; more than 700 of them have been specifically identified in U.S. drinking water supplies. The technology for removing most of them is known but rarely applied.

Water quality standards were established by the Safe Drinking Water Act of 1974, which required the USEPA to set minimum standards. Only a limited number of criteria have been set to date. The Safe Drinking Water Act Amendments of 1986 set up an ambitious schedule to implement new or revised regulations over the five-year period 1987–1991. Enforcement by the states must start within a fixed time frame. A broad range of treatment methods is available to obtain better water quality. Treatment is generally applied as close to the point of use as possible, while still serving its intended purpose.

Sand filtration is a widely accepted process for removing suspended solids. Since this is a simple method requiring little technology, it is widely used in developing countries. Activated carbon filters improve both the taste and odor of drinking water. Dosing with a flocculant such as alum improves settling and filtration of suspended matter in storage tanks. In reverse osmosis, water is passed through a membrane to remove dissolved matter (organic, minerals, gas). Reverse osmosis is also widely used for seawater desalination in countries that lack surface or groundwater supplies.

To reduce pipe corrosion, lime is added to water of certain quality. Fluorides are added to water supplies to improve dental health. Chlorine is added to almost all water supplies for disinfection and to kill any organisms remaining after all other treatment.

## COSTS

The cost of constructing water supply systems is generally borne by localities and by the states. The cost of Tunnel No. 3, built to supplement the water supply delivery capacity for New York City, was $60 million per mile (Mc-Gough, 1983). Replacement of large trunk mains costs approximately $1000 per foot, and smaller distribution mains are often estimated at $110 per foot. The cost of cleaning and cement lining of water mains has been estimated at $56 per foot (McGough, 1984). Revenue is collected to pay for some of these improvements. The cost of water to the user averages $1 per 1000 gallons in the United States as a whole. The cost in New York City is $0.80 per 1000 gallons. The sewerage charge, which is related to the consumption, is an additional 30% of the water supply cost.

## METERING

Almost all new construction includes metering of the water supply. In New York City, many buildings were built before there were metering requirements; these buildings are charged on the basis of feet of frontage or according to the number of outlets (showers, faucets, etc.) on the premises. This is clearly not conducive to water conservation, and rules to require meters for all buildings are now in effect.

## PUBLIC VERSUS PRIVATE OWNERSHIP

In the United States, 82% of urban water systems are publicly owned, although there are some large private systems, regulated by public service commissions. Whether or not a public system is justifiable depends on the settlement density of the area. The following criteria will serve as a guide (U.S. Public Health Service, 1962):

| | | |
|---|---|---|
| Over 2500 persons per square mile | Lots of less than 1-acre | Public system always justified |
| 1000 to 2500 persons per square mile | Lots of 1 to 2 acres | Public system normally justified |
| 500 to 1000 persons per square mile | Lots of 2 to 4 acres | Public system not normally justified |
| Fewer than 500 persons per square mile | Lots of more than 4 acres | Public system rarely justified |

## PLANNING CONSIDERATIONS

Urban water supplies are under the control of building, public works, and health departments. Rural water supply, where it is not obtained from on-site wells, is controlled by water rights that run in conjunction with the land. These rights merely assure the downstream landowner that the upstream landowner will not interfere with the flow of water before it reaches the boundary of his or her property.

Lack of water not only limits human settlement, but can seriously interfere with industry. The processing of oil shale planned for the western slope of the Rocky Mountains in Colorado, which would require vast quantities of water for washing, was severely hampered by the shortage of water. Conversely, water supply restrictions can be used effectively as growth controls. The city of Boulder, Colorado refused to extend its water supply mains outside the city's boundaries in the hope of setting a limit on sprawling urbanization.

Management of this important resource is the most effective planning tool. Needs must be assessed continually, new supplies must be sought out, and up-to-date technology must be applied. Preventive maintenance must be planned for and adequately funded. Rehabilitation of older parts of the system must be regularly scheduled. Fees must be collected to pay for the upkeep. A well-managed system will supply good-quality water in adequate quantity at reasonable cost in an unobtrusive manner.

If properly planned for, *gray water* from a subdivision can be reused after minimal treatment. Gray water is the effluent from showers, laundries, or once-through cooling processes, which contains no biological wastes and therefore can be reused for irrigation, fire protection, or even for the flushing of toilets and urinals. It is necessary to have carefully separated piping systems for this purpose, but this is a small price to pay for useful water in water-short areas. A blue dye is usually added to gray water so that inadvertent cross connections can be promptly identified.

Ponds and lakes should be designed for multiple uses. A pond can serve as a retention basin for storm water, to prevent floods in an area. The pond can serve as a recharge basin for an aquifer that is being drawn on too heavily. Solar energy storage can be achieved with a carefully designed pond. Finally, water can be used for its aesthetic value. Where fountains are included, ponds can substitute for unsightly cooling towers. When recreational uses for the pond are added, the best of all multiple purposes has been attained.

## Planning for Conservation

Based on the land-use capacity as expressed in the master plan for a region, the future requirements for water supply can be estimated. The water resources of the region can be determined from hydrological data available from local governmental sources, or from hydrogeologists working in the area. A comparison of these data will show clearly whether the region will be able to support the planned development.

Water shortages can usually be predicted if full development is allowed to occur. A region then has the choice of developing additional capacity (if it can be found) at a cost. Where shortages are seasonal or the costs too high, communities may choose not to fund such expenditures. The alternative is to establish a policy that water conservation is to be planned for; regulations must be promulgated and enforced. The local land-use planner can be very influential. Whenever a subdivision or a large development is submitted for approval, the planner must not assume that water will be available automatically; potential consumption data must be submitted for review to determine how the supply will be made available from the region's resources.

Recommendations and regulations for water conservation must include the use of water-saving fixtures, metering, denser settlement, and water-conserving landscaping. The use of drought-tolerant plants is being fostered in many western states that regularly experience seasonal droughts. Education aimed at making the public more aware of water conservation opportunities is an integral part of any conservation program and can be carried out by a number of local government departments, including education, public works, finance (by setting rates), and planning. Setting water rates at the full cost of maintaining and improving an area's water supply system has been found to be the most effective method of creating conservation attitudes.

## REPAIR AND REHABILITATION

Inner-city water supply systems started at the turn of the century and still in service have exceeded their intended life cycles. On the average, 50 breaks per year occur per 1000 miles of pipe. Given budgetary constraints, it would be impossible to replace entire distribution systems at one time. However, the older cities on the east coast of the United States, where the oldest water systems exist, have instituted multiyear replacement programs. In the meantime, sections are replaced as they fail, causing necessary inconvenience.

As we note in Chapter 12, 60% of funding required for water supply infrastructure in the years to come will go toward rehabilitation and repair. It is essential that the balance of the funds that will be used for system expansion be supplemented to provide for preventive maintenance; otherwise, the systems will continue to deteriorate at an accelerating rate.

## ORGANIZATIONS

Notable organizations in the water supply field include the following:

American Public Works Association
American Water Works Association
Water Pollution Control Federation
National Board of Fire Underwriters (fire protection standards)
U.S. Public Health Service (water quality standards)

## REFERENCES

Cohen, P., O. L. Franke, and B. L. Foxworthy (U.S. Geological Survey), *An Atlas of Long Island Water Resources*, State of New York, Albany, 1968.

Congressional Budget Office, *Public Works Infrastructure: Policy Considerations for the 1980's*, U.S. Government Printing Office, Washington, D.C., 1983.

McGough, J. T., *The Third Water Tunnel*, NYC Department of Environmental Protection, New York, 1983.

McGough, J. T., *Aging Infrastructure: A Need for Federal Funding of Urban Water Supply Systems*, NYC Department of Environmental Protection, New York, 1984.

Public Health Service, *Environmental Health Planning Guide*, U.S. Department of HEW, Washington, D.C., 1962.

Solley, Wayne B., Charles F. Merk, and Robert Pierce, *Estimated Use of Water in the United States in 1985*, U.S. Geological Survey Circular 1004, U.S. Government Printing Office, Washington, D.C., 1988.

Swenson, H. A., and H. L. Baldwin, *Primer on Water Quality*, U.S. Department of Interior, Washington, D.C., 1965.

## ADDITIONAL READING

Baldwin, H. L., and C. L. McGuiness, *Primer on Ground Water*, U.S. Department of Interior, Washington, D.C., 1966.

Barcelona, M., J. F. Keeley, W. A. Peetyjohn, and Allan Wehrman, *New Handbook on Groundwater Protection*, Hemisphere Publishing, New York, 1988.

Culp, Wesner, and Culp, Inc., eds., *Public Water Systems*, Van Nostrand Reinhold, New York, 1986.

Jaffe, Martin, and Frank DiNovo, *Local Groundwater Protection*, APA Planners Press, Chicago, 1987.

Wade Miller Associates, Inc., *Water Supply*, National Council on Public Works Improvement, Washington, D.C., 1987.

Sanders, W., and Charles Thurow, *Water Conservation in Residential Development: Land-Use Techniques*, APA Planning Advisory Service 373, APA Planners Press, 1982.

## QUESTIONS

1. Describe the hydrological water cycle as it occurs in nature.

2. What is the largest segment of water consumption in the United States?

3. What are two main sources of water? In what proportion are they found in the United States?

4. How is water pressure created?

5. What is the range of daily water consumption for a town with a population of 30,000 people? What is the average cost per person?

6. Name at least two factors that influence water quality. Give the range of each and the units in which it is measured.

7. How can good urban design and effective planning foster conservation of water consumption?

8. What federal regulations control potable water?

9. What treatment is applied to almost all water supplies for disinfection purposes?

10. How are water mains protected from freezing?

Arthur Kill Water Pipe Installation

# 2

# SEWERS AND STORM DRAINAGE

**George Rainer**
Flack & Kurtz Consulting Engineers

## SEWER SYSTEM TYPES

There are three distinct types of sewers: sanitary, storm, and combined. *Sanitary sewers* carry away the wastewater from residential and institutional uses, frequently combined with industrial effluents. *Storm sewers* carry the runoff from rain which is collected from roofs, roads, and other surfaces. *Combined sewers* carry both sanitary sewage and storm water.

In drainage systems started during the second half of this century, sanitary and storm drainage are generally kept separate. The major reason for this separation is the fact that sanitary sewers generally discharge to sewage treatment plants, which must be designed to handle the maximum flow of the sewer system. During rainstorms the flows of combined systems can swell to as much as 50 times the normal flows; this either requires that the treatment plant be oversized, or causes untreated sewage to spill into waterways.

Where sewer systems were originally designed as combined sewers (as in New York City), separation becomes costly and can occur only over extended periods. Separate sewers in New York City now comprise 40% of the total, whereas the fraction is 98.4% in Westchester County and 100% in Nassau and Rockland counties in the New York metropolitan area. Regulators or diversion gates in the combined system limit the flow going to treatment plants during intense rainfall to twice the dry weather flow, while spilling the excess rainwater to waterways.

## SEWER CONFIGURATIONS

Sewer systems consist of a network of pipes that are generally buried several feet below the street level. The connections from individual sites, called *laterals*, terminate in submains; these in turn empty into trunk sewers, which lead to a treatment plant. After treatment the water is discharged through an outfall sewer to a nearby waterway.

Wherever possible, sewage flows by gravity; pipes are pitched in a downward direction at a rate of 0.5 to 2%, but always steeply enough to maintain a velocity of 2 feet per second, which is the minimum required for self-cleaning. A maximum of 10 feet per second is usually designed for to avoid any erosion in the piping that might result from the entrainment of grit in the sewage.

In the design of sewers it is intended that the pipes should flow half-full; this permits ventilation of the pipes. When a gravity sewer normally flows full, it is considered to be a pressure conduit and is a sign of inadequate capacity; during periods of slight excess flow, then, water will fill up manholes, which is known as *surcharging*.

When an area is being developed at a great distance from the main section of a piped sewer system, or where the local terrain requires it, sewage can be collected at a pumping station and conveyed to the sewage treatment plant through a force main. This is generally not desirable because pumping stations require maintenance and use energy, but sometimes it is unavoidable.

Sewage leaves a building through a cast-iron house drain provided with a vented trap that keeps odors out of the building's interior. Sewer pipes up to 18 inches in diameter are usually constructed of clay pipe; sizes larger than that are generally fabricated of concrete with gasketed joints. Plastic pipe is becoming more widely accepted in sewer construction, due to its lighter weight and ease of joining.

When installed on unstable soil such as a former landfill, clay, or silt, sewers must be supported on concrete cradles to avoid settlement and cracking. Where sewers change direction, where several runs come together, and at specified intervals on straight runs, *manholes* are provided. These are poured concrete or prefabricated structures with ladders and steel covers that permit access for inspection, cleaning, and repair.

## SOURCES AND QUANTITIES OF SEWAGE

The quantities of sewage generated are clearly related to the amounts of water supplied to a given facility. The figure "100 to 200 gallons per person per day" cited in Chapter 1 for water supply requirements included all functions likely to be engaged in by a "statistical person." Table 2-1 disaggregates these daily consumption figures into specific functions, including domestic, institutional, commercial, and industrial wastewater flows. In planning a sewage system for a given development area, the zoning map will show the existing land uses and the permitted density of settlement. This will determine the total population to be expected in the area and will permit sewer design to go forward.

Not all the water supplied to an area will leave as sewage: Water is used for pools, sprinklers, cooling towers, and so on, some of which evaporates to the atmosphere; and some is lost through leaks in the supply piping. Nevertheless, the quantities going to treatment plants often exceed the water supply. This is caused by infiltration into the sewer system from groundwater and tidal inflow. In New York City, for example, 1300 million gallons of water is supplied each day, but the treatment plants handle 1536 million gallons per day.

## STORM WATER DRAINAGE DESIGN

Urban storm water runoff depends on the local rainfall pattern and the type of surface being drained. The basic formula is

$$Q = CIA$$

**TABLE 2-1   ESTIMATED SEWAGE FLOWS**

| Building Use | Range (gallons per capita per day) |
|---|---|
| School (8-hour day) | |
|    Without cafeterias, gyms, or showers | 8–10 |
|    Without cafeterias | 12–15 |
|    With cafeterias, gyms, and showers | 20 |
| Housing | |
|    Small dwellings | 50 |
|    Large dwellings | 75–100 |
|    Multiple family residence | 60 |
|    Trailers with individual baths | 50 |
| Commercial | |
|    Hotel kitchens, three meals/day | 7–10 |
|    Restaurants | 7–10 gal/patron |
|    Motels | 25–40 |
|    Hotels | 100 |
| Industrial (not including industrial wastes) | 15–35 gal/person/shift |

where

$Q$ = quantity, cubic feet per second

$C$ = coefficient of runoff: paved areas, 0.9; grass areas, 0.3; roof areas, 1.0

$I$ = intensity of rainfall, in./hr (Table 2-2)

$A$ = area drained, acres (1 acre = 43,560 sq ft)

The velocity to be maintained in storm drains is 5 feet per second, in order to carry along gravel and debris usually contained in stormwater. *Catch basins* (concrete pits with gratings on top) should be placed every 300 to 600 feet.

It is very important to incorporate storm drainage into green space planning rather than to allow the problem to be solved through engineering means alone. Ponds properly located can serve as retention basins to delay runoff after a heavy rainfall. As this runoff is delayed, impurities will settle out, sending cleaner water downstream. Where the soil is porous enough, the pond will act as a recharge basin for groundwater supplies. In reshaping the contours of the land

**TABLE 2-2   SELECTED RAINFALL REFERENCE DATA**

| Location | Inches/ Peak Hour | Location | Inches/ Peak Hour |
|---|---|---|---|
| Alabama, Birmingham | 3.6 | Montana, Billings | 2.0 |
| Alaska, Anchorage | 1.0 | Nebraska, Lincoln | 3.6 |
| Arizona, Phoenix | 2.5 | Nevada, Reno | 1.2 |
| Arkansas, Little Rock | 3.7 | New Hampshire, Concord | 2.4 |
| California, Los Angeles | 2.0 | New Jersey, Newark | 3.0 |
| California, San Francisco | 1.6 | New Mexico, Albuquerque | 2.1 |
| Colorado, Denver | 2.5 | New York, Buffalo | 2.4 |
| Connecticut, Hartford | 3.0 | New York, New York | 3.2 |
| Delaware, Wilmington | 3.5 | North Carolina, Raleigh | 3.7 |
| D.C., Washington | 3.4 | North Dakota, Bismark | 2.4 |
| Florida, Miami | 4.6 | Ohio, Cincinnati | 2.7 |
| Georgia, Atlanta | 3.5 | Oklahoma, Tulsa | 3.9 |
| Hawaii, Honolulu | 3.2 | Oregon, Portland | 1.4 |
| Idaho, Boise | 1.1 | Pennsylvania, Philadelphia | 3.3 |
| Illinois, Chicago | 3.0 | Puerto Rico, San Juan | 4.0 |
| Indiana, Evansville | 3.0 | Rhode Island, Providence | 3.0 |
| Iowa, Des Moines | 3.3 | South Carolina, Charleston | 4.1 |
| Kansas, Topeka | 3.7 | South Dakota, Rapid City | 2.7 |
| Kentucky, Lexington | 2.8 | Tennessee, Nashville | 3.1 |
| Louisiana, New Orleans | 4.5 | Texas, Houston | 4.5 |
| Maine, Portland | 2.3 | Utah, Salt Lake City | 1.3 |
| Maryland, Baltimore | 3.4 | Vermont, Burlington | 2.0 |
| Massachusetts, Boston | 2.6 | Virginia, Richmond | 3.6 |
| Michigan, Detroit | 3.0 | Washington, Seattle | 1.0 |
| Minnesota, Minneapolis | 3.0 | West Virginia, Charleston | 2.8 |
| Mississippi, Jackson | 3.8 | Wisconsin, Madison | 3.0 |
| Missouri, St. Louis | 3.4 | Wyoming, Cheyenne | 2.5 |

*Source: Historical Climatological Series, Precipitation No. 4-2, NOAA, Ashville, N.C., 1987.*

in a large development it is essential to respect the natural drainage channels of the site. As can be seen from the runoff coefficients listed above, paved areas have much larger runoffs than those of grass areas, and this should be considered carefully in developing site plans.

Flooding and drainage are always local problems. The federal government does not even set guidelines as to whether drainage design should consider rainfall that recurs statistically every 5, 10, or 100 years. The more conservative the design, the more costly the solution; the trade-off between spending public funds and gambling on safety must be made at the local government level.

## QUALITY OF STORM WATER

Although storm sewers carrying runoff only are now generally discharged directly into waterways, this practice is being questioned more and more frequently. Table 2-3 shows a comparison between the quality of urban storm water runoff and the overflow from a combined sewer. The chlorides in storm water clearly come from highway deicing efforts; oils and lead are also related to automobile traffic; and solids of all types are prevalent in greater quantities. Clearly, storm water is far from clean and a movement to require at least some form of primary treatment for storm water alone is gaining momentum.

## WASTEWATER TREATMENT METHODS

Where settlement is fairly sparse (2500 persons per square mile or less), on-site sewage treatment is usually possible. In the earliest days of sanitation this consisted of a hole in the ground with an outhouse over it. The wastes were dosed with lime periodically, and ultimately, a new hole was dug. Nowadays this is accomplished by installing septic tanks with drainage fields (Fig. 2-1).

**TABLE 2-3   COMPARATIVE CHARACTERISTICS OF STORM WATER RANGE OF VALUES**

| Characteristics | Urban Storm Water | Combined Sewer Overflows |
|---|---|---|
| Biological/oxygen demand (mg/L) | 1–700 | 30–600 |
| Chemical oxygen demand (mg/L) | 5–3100 | — |
| Total suspended solids (mg/L) | 2–11,300 | 20–1700 |
| Total solids (mg/L) | 450–14,600 | 150–2300 |
| Volatile total solids (mg/L) | 12–1600 | 15–820 |
| pH | — | 4.9–8.7 |
| Settleable solids (mg/L) | 0.5–5400 | 2–1550 |
| Organic nitrogen (mg/L) | 0.1–16 | 1.5–33.1 |
| $NH_3N$ (mg/L) | 0.1–2.5 | 0.1–12.5 |
| Soluble $PO_4$ (mg/L) | 0.1–10 | 0.1–6.2 |
| Total $PO_4$ (mg/L) | 0.1–125 | — |
| Chlorides (mg/L) | 2–25,000 | — |
| Oils (mg/L) | 0–110 | — |
| Phenol (mg/L) | 0–0.2 | — |
| Lead (m/L) | 0–1.9 | — |
| Total coliforms (number/100 ml) | 200–146 × 10⁶ | 20,000–90 × 10⁶ |
| Fecal coliforms (number/100 ml) | 55–112 × 10⁶ | 20,000–17 × 10⁶ |
| Fecal streptococci (number/100 ml) | 200–1.2 × 10⁶ | 20,000–2 × 10⁶ |

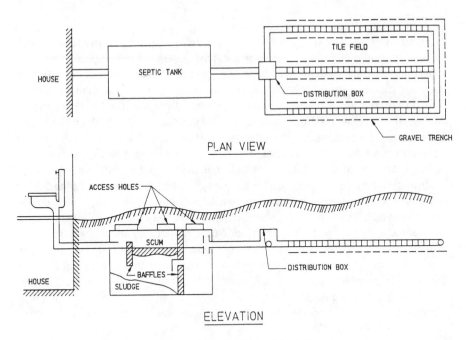

**Figure 2-1** Septic tank layout.

The size of these installations depends on the capability of the soil to absorb the effluent. Once percolation capability has been established by performing a simple test, system sizing can proceed according to well-defined engineering standards (Table 2-4).

Effluent is discharged into a concrete pit, generally sized to hold sewage for

**TABLE 2-4   SEPTIC TANK ADSORPTION FIELD**

| Percolation Rate (in./min) | Required Area of Adsorption Field, per Bedroom (sq ft) |
|---|---|
| Greater than 1 | 70 |
| Between 1 and 0.5 | 85 |
| Between 0.5 and 0.2 | 125 |
| Between 0.2 and 0.07 | 190 |
| Between 0.07 and 0.03 | 250 |
| Less than 0.03 | Unsuitable ground |

*Source:* Vesilind and Peirce (1980).

24 hours; partial decomposition occurs during that time, as organic matter is digested through anaerobic action. Solid particles settle out in the tank, which requires periodic cleaning. The overflow passes into a field of perforated tile, which allows the liquid to seep into the ground. It is important to maintain adequate distance from groundwater sources in designing these drainage fields.

In urbanized areas treatment occurs at central plants. The major elements that must be removed from sewage are suspended solids, biological oxygen demand (BOD), nitrogen, phosphorus, and coliform bacteria. The level of treatment provided is closely related to its cost; purifying the effluent to a level higher than that of the stream to which it is discharged is unnecessary.

Primary treatment is provided for all sewage entering a treatment plant. This consists of coarse *bar screens* that remove large items of floating material; the screens are raked automatically and the material is disposed of. The next step is a *comminutor*, which is a circular grinder that reduces to smaller particles the material that has passed through the bar screens. This is followed by a *grit chamber*, where the velocity of approach is reduced to 1 foot per second to allow the heavier particles (e.g., grit and sand) to settle out. The grit must be removed to prevent damage to pumps and other equipment during subsequent processes; it can be removed to landfill without any danger of odor problems (Vesilind and Peirce, 1980).

The next step in most treatment plants is the *settling tank*, or *primary clarifier*, where turbulence is kept to a minimum and retention time is made as long as possible. (Retention time is the total time a quantity of water will spend in the tank, typically 45 minutes to 2 hours.) The organic matter, which floats to the top is called *scum*, and the solids, which settle to the bottom of the tank, are known as *raw sludge*, which is removed by scrapers through a pipe into a sludge tank. As rotating skimmers in the primary clarifier move across the surface of the water to remove grease, the clarified liquid escapes through a weir at the top of the tank. Primary treatment can remove from 50 to 60% of the suspended solids and can reduce the BOD content by 30 to 50%.

Water leaving the primary clarifier has lost much of the organic matter originally contained, but it retains a high oxygen demand, which means that it is composed of molecules that will decompose through microbial action in the presence of oxygen. Whereas it was the purpose of the primary treatment to remove the solids, it is the purpose of the secondary treatment to reduce the need for oxygen, or to reduce the BOD. This is generally accomplished by facilitating the microbial action of the microorganisms through a variety of methods.

In a *trickling filter* a rotating arm spreads the liquid waste over a filter bed consisting of fist-sized rocks. A very active biological growth forms on the rocks, referred to as *slimes*, and the microorganisms obtain their food from the waste stream trickling over the rocks. Air passes through the rocks either by natural circulation or by forced circulation, to speed the process. The effluent flows through a final sedimentation tank to remove the dead organic matter from the filter stones and any final particles that have escaped the filter. Trickling filters are sensitive to very cold climates and to industrial effluents.

The more prevalent secondary treatment method is the *activated sludge system*. Effluent from the primary clarifier is brought to an aeration tank full of microorganisms. Air is bubbled through this tank to provide the oxygen necessary for the survival of these aerobic organisms. The microorganisms come into contact with the dissolved organics and rapidly absorb them on their surface. In time the microorganisms decompose the material to $CO_2$, $H_2O$, some stable compounds, and more microorganisms. The production of these organisms is relatively slow, taking from 4 to 8 hours. The sewage then passes to the final settling tank, where it resides for 1 or 2 hours. More than half of the sludge that settles out here goes to sludge holding tanks; the balance is returned to the aeration tank, where it is used to speed the decomposition process. Effluent from the final settling tank goes to a waterway or to tertiary treatment.

Secondary treatment reduces suspended solids by 85 to 95%, BOD by 80 to 95%, and coliforms by 90 to 95%. This is a very efficient process, and all sewage treatment plants being built today provide secondary treatment. Of the 15,438 treatment plants in existence in the United States in 1986, only 15.1% provided less than secondary treatment (Apogee Research, Inc., 1987) (Fig. 2-2).

Where the receiving waterways are very sensitive, where specific pollutants are prevalent and must be removed, or where funding is readily available, tertiary treatment is provided for sewage. This can be in the form of a micro-strainer (for solids removal), oxidation ponds (for BOD removal), activated carbon adsorption (removes inorganics as well as organics), denitrification, alum dosing (phosphate removal), or any number of physiochemical processes. In addition to tackling specific pollutants, tertiary treatment results in BOD removal of 98 to 99%.

Industrial wastewater frequently requires special treatment. Dissolved metals present a particularly difficult problem because they are costly to remove. Although such specialized treatment can be instituted at the central sewage treatment plant, it is generally more practical to provide it at the source—at the industrial user's site. This method also relegates the cost of special treatment to the source of the problem. Where federal grants have been used to build a public sewage treatment plant, local government is required to assess the industrial user for the special treatment required. Where a large development is built out of range of existing sewer lines, it is possible to provide small packaged sewage treatment plants. Such plants provide secondary treatment but require a fair amount of maintenance. The final effluent from any plant is generally chlorinated for odor control and disinfection, or it may be treated with ozone or ultraviolet light.

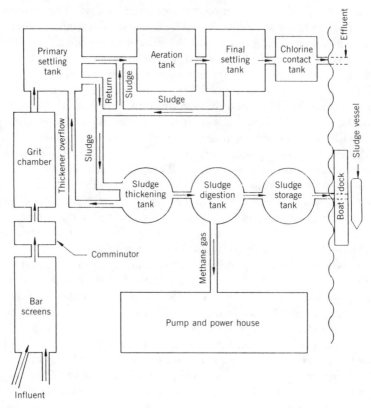

**Figure 2-2** Water pollution control plant, secondary treatment. (From NYC Department of Environmental Protection.)

Instead of using expensive tertiary treatment, it is possible to spray secondary effluent on land, where the soil microorganisms can degrade the remaining organics. Certain crops irrigated in this manner will benefit from the nutrients (e.g., nitrogen, phosphate) still present in the wastewater.

## SLUDGE DISPOSAL

Since sludge generally consists 95% of water, it must be treated before disposal. To reduce treatment and transportation costs, the sludge is placed in a settling tank called a *thickener.* After settling for a period of 12 to 24 hours, a large amount of water separates from the sludge and is returned to the beginning of the process.

Anaerobic digestion takes place in the digester, where the remaining sediment from primary or secondary clarification has been collected. This process causes a 50 to 60% reduction in volatile solids; for each pound destroyed, 15 to 18 cubic feet of gas with a heat content of 600 Btu per cubic foot are produced. This gas, which is rich in methane, can be used for process and comfort heating but is frequently flared off.

The digestion process will occur naturally over time, but it can be speeded up if the digester temperature is kept at 100°F and an alkali pH is maintained while fresh sludge is slowly added. The digested sludge must then be dried, either in large open beds (1 to 2 square feet per capita served) or by means of a vacuum filter. The dried sludge is disposed of by dumping at sea, by incineration, or is used as landfill.

## PLANNING CONSIDERATIONS

Whereas it is a straightforward task to determine the size of a sewer system based on the zoning capacity of the area it is to serve, it is much more difficult to size a sewage treatment plant, because of the time factors involved. A treatment plant designed for the ultimate capacity of the area it will serve is bound to be oversized for its initial service. This is wasteful of construction dollars and may require operating modifications to make it work efficiently at partial flow.

As part of an effort to clean up the nation's waters, the federal Environmental Protection Agency (EPA) initiated a construction grants program for the building of water pollution control plants. Some of the recipients of these grants built plants that would not be fully used for decades to come. This short-changed areas that were in dire need of funds for additional sewage treatment capacity in the near term. This led to a requirement for the submission of comprehensive land-use plans with the application for new treatment plants funds. On the other hand, the declaration of building moratoria caused by the lack of treatment plant capacity has become a familiar development issue. In these instances, planning was inadequate and infrastructure was simply not available when development was ready to proceed based on population pressure or economic opportunity.

Siting a new sewage treatment plant can be a problem. It must be located in the vicinity of a receiving water course. It should be located geographically at the lowest point in the area it is to serve, to minimize the need for pumping. Preferably, it should be located in an area remote from residential uses because of its visual impact and the possible existence of odors during periods of malfunction. The area required for a plant capacity of 50 million gallons per day (MGD) is 12.5 acres. Although these requirements can cause strenuous opposition from prospective neighbors of such a plant, such facilities are an absolute necessity for urbanized areas, and local government must minimize their impact to get them built.

## REGULATORY CONSIDERATIONS

Although a number of minor water pollution control measures have been passed by Congress since the beginning of the twentieth century, it was not until 1948 that a comprehensive measure, P.L. 80-845, was adopted. This act placed the Surgeon General of the Public Health Service and the Federal Works Administrator in charge of implementing its provisions. A comprehensive program to eliminate pollution was to be prepared, technical information was to be collected and disseminated to the states; and states were encouraged to form interstate compacts to deal with pollution problems more effectively. Enforcement procedures were quite weak and took an inordinate amount of time to complete. Some funding was made available.

Over the years the act was amended a number of times; responsibility was shifted to different departments; funding was increased, enforcement authority was strengthened, and jurisdiction was extended. Finally, in 1970, the National Environmental Protection Act (P.L. 91-190) became effective, and EPA was established. EPA took over the pollution control functions of a wide variety of departments and agencies, but the important impact on water pollution control was its assumption of the administration of the Federal Water Pollution Control Act of 1972 (P.L. 92-500).

The 1972 Act replaced P.L. 80-845 and set as its goal "to restore and maintain the chemical, physical and biological integrity of the Nation's waters." Instead of merely assisting the states in managing water quality, the act required the federal government to take a lead in the pollution control effort, with assistance from the states. EPA was made responsible for setting water quality standards, developing criteria, establishing effluent limits, and developing a National Pollution Discharge Elimination System (NPDES), which issues permits to all municipalities and industries that discharge wastewater. Federal financing was also greatly increased under the act through the construction grants program, which was the most effective catalyst.

The Clean Water Act of 1977 (P.L. 95-217) required all of these programs to be delegated to the states, with oversight from EPA, as soon as they had acquired the ability to do so. The major innovation of this act was its section 208, which called for the development of areawide waste treatment management plans. These plans, which were funded throughout the nation, became the most important source of information for assessing the budgetary needs of our sewer infrastructure.

Under the most recent federal legislation, the Water Quality Act of 1987 (P.L. 100-4) will give the operation of the entire water pollution control program to the states by 1994. The construction grants will be replaced by state revolving funds, and the federal share will be eliminated. However, enforcement will continue to be guided by the rules set up by EPA. It is evident from the legislative trends that the states will have to assume much greater financial and administrative responsibilities in the wastewater area in the years to come.

## COSTS

Capital costs for wastewater treatment plants are considerable. A plant handling 50 MGD and providing secondary treatment will cost $150 million in 1987 dollars. Funds for operation and maintenance of a typical plant can be estimated at $37 per year per person served (1984 dollars) (Apogee Research, Inc., 1987). Typical costs for sewer system components, including excavation and backfill, are as follows:

| | |
|---|---|
| Manhole or catch basin | $1200 each |
| 3-inch ductile iron pipe | $10 per foot |
| 6-inch ductile iron pipe | $40 per foot |
| 8-inch vitrified clay pipe | $60 per foot |
| 15-inch reinforced concrete pipe | $80 per foot |
| 18-inch reinforced concrete pipe | $90 per foot |
| 6-inch gate valve in box | $514 each |

Operating costs are generally met out of user charges. As indicated in Chapter 1, sewer charges are generally 25 to 30% of water supply costs. As treatment charges escalate, user charges will increase proportionally. Research into more economical methods of treatment should clearly be undertaken.

## FUTURE REQUIREMENTS

Like any other piping system, underground sewer systems will deteriorate over time, and the costs of repair must be budgeted for. However, in a needs assessment prepared by EPA for the year 2006, it appears that only 3.9% of funds are earmarked for replacement and rehabilitation, with another 3.4% going to correct inflow and infiltration problems. The balance of the needs deal with improving treatment plants (36.8%), reducing combined sewer overflows (19.8%), and providing new collector and interceptor sewers to meet growing needs (36.1%). It appears, therefore, that repair of the sewer infrastructure is less of a problem than are improvement and extension of facilities. The needs assessment does not take into account the cost of maintenance and operation, and that will have to be the area of major expenditures in the years to come.

## REFERENCES

Apogee Research, Inc., *Waste Water Management*, National Council on Public Works Improvement, Washington, D.C., 1987.

Vesilind, P. Arne, and J. J. Peirce, *Environmental Pollution and Control*, Ann Arbor Science Publishers, Ann Arbor, Mich., 1980.

## ADDITIONAL READING

Abbet, R. W., *American Civil Engineering Practice*, Vol. 2, Wiley, New York, 1957.

Hazen and Sawyer, *Section 208 Area Waste Treatment Management Program*, NYC Department of Environmental Protection, New York, 1979.

Nelson, Stephen C., *Interrelationships of Land Use and Water Quality*, Voorhees & Associates, McLean, Va., 1973.

NYC Department of Environmental Protection, *An Overview of the Department's Responsibilities and Functions*, New York, 1982.

Arthur Young, Metcalf and Eddy, *Municipal Wastewater Management: Citizen's Guide to Facility Planning*, NYC-EPA FRD-6, NYC Environmental Protection Agency, New York, 1985.

## QUESTIONS

1. Define the types of sewer systems in use.

2. What is the principal disadvantage of combined sewers?

3. What are the minimum and maximum velocities in sanitary sewer pipes, and why?

4. How much wastewater (in gallons per day) is generated by a school with 300 pupils that has a cafeteria and a gym with showers?

5. What is the major method used for secondary sewage treatment? What is the major factor being improved, and by how much?

6. Which legislation fostered areawide planning for water pollution control?

7. Calculate the stormwater drainage from the following development in Phil-

adelphia: 2.5 acres of grass, 1 acre of paved areas, and 10 buildings with roofs of 1000 square feet each.

8. What financing mechanisms are proposed for the states in the future?

9. Describe the most basic method of sewage treatment in sparsely populated areas.

10. What is the reason for requiring that sewage treatment plant capacity be greater than the water supplied to an area?

Resource Recovery Plant (Peekskill, New York)

# 3

# SOLID AND HAZARDOUS WASTES

**George Rainer**
Flack & Kurtz Consulting Engineers

# SOLID WASTES

Solid waste management, which includes the collection, transportation, treatment, and disposal of wastes, is a crucial element of the city's infrastructure. Although the collection of solid wastes is always a local service problem and the disposal issue is generally regional in nature, the broad policy issues must be dealt with on the federal level.

## Types of Waste

Solid wastes can be broadly classified in the following categories:

Garbage
Rubbish
Ashes
Street refuse
Building wastes
Industrial wastes

*Garbage* consists of wastes generated by the preparation, cooking, and serving of food, as well as from the handling, storage, and sale of produce; it is characterized by a high moisture content. The combustible fraction of *rubbish* consists of items such as paper, cartons, boxes, barrels, tree branches, and furniture. The noncombustible fraction consists of metal objects, tin cans, glass, crockery, minerals, and the like. *Ashes* are the residues from fires used for heating and from incineration.

*Street refuse* includes sweepings, dirt, leaves, catch basin residue, and the contents of litter receptacles. Other items found in streets and on sidewalks are dead animals, abandoned cars, and horticultural wastes. *Building wastes* include scrap lumber, pipe, and other materials left over from construction, as well as items generated from the demolition of buildings or the repair of roads and other infrastructure. *Industrial wastes* include a broad variety of substances, including hazardous materials from commercial production lines, pathological wastes from hospitals, or radioactive materials from power plants. Sewage treatment plant residue, including solids from coarse screening and grit chambers, and the final sludge, are part of this waste stream.

Although the composition of wastes from specific sites will vary widely depending on land use, a breakdown for a heavily urbanized area would typically be as follows:

| | |
|---|---|
| Household wastes | 48% |
| Commercial | 31% |
| Construction | 5% |
| Other | 16% |

## Quantities

To gain a perspective on the solid waste problem, it must be noted that 450,000 tons of solid wastes are generated in the United States each day (R. W. Beck and Associates, 1987); of this quantity 28,000 tons per day are created in New York City alone (NYC Department of Sanitation, 1982). Individually, each person generates an average of 4 pounds of solid wastes per day. Other unit factors can be found in Table 3-1. Table 3-2 is a survey of industrial waste quantities generated by a number of industries listed by Standard Industrial Code number.

## Collection Methods

Within buildings, wastes are collected by means of chutes, hoists, or elevators. In high-rise office buildings the waste is bagged on a floor-by-floor basis and brought down to a loading dock by elevator. High-rise apartment buildings and hospitals generally utilize chutes to bring wastes to a central service area on the lowest floor. Where manned or unmanned hoists are available for the movement of materials, as in the case of factories or construction projects, they are generally used for the removal of wastes.

### TABLE 3-1  COMMUNITY SOLID WASTE GENERATION

| Building Use | Waste Production |
|---|---|
| Residential | 4 lb/person/day |
| School | |
| K–6 | 10 lb/room plus 0.5 lb/pupil |
| 7–12 | 8 lb/room plus 0.5 lb/pupil |
| College | 4 lb/100 sq ft/day |
| Community facility | 1 lb/100 sq ft/day |
| Health Facility | 2 lb/100 sq ft/day |
| Commercial | |
| Office | 1 lb/100 sq ft/day |
| Retail | 10 lb/100 sq ft/day |
| Theater | 0.25 lb/person |

*Source:* Flack and Kurtz Consulting Engineers (1974).

**TABLE 3-2   INDUSTRIAL SOLID WASTE GENERATION**

| SIC Code | Industry | Waste Production Rate (tons/employee/year) |
|---|---|---|
| 2010 | Meat processing | 6.2 |
| 2033 | Cannery | 55.6 |
| 2037 | Frozen foods | 18.3 |
| Other 203 | Preserved foods | 12.9 |
| Other 20 | Food processing | 5.8 |
| 22 | Textile mill products | 0.26 |
| 23 | Apparel | 0.31 |
| 2421 | Sawmill and planing mills | 162.0 |
| Other 24 | Wood products | 10.3 |
| 25 | Furniture | 0.52 |
| 26 | Paper and allied products | 2.00 |
| 27 | Printing and publishing | 0.49 |
| 2810 | Basic chemicals | 10.00 |
| Other 28 | Chemicals and allied products | 0.63 |
| 2900 | Petroleum | 14.8 |
| 3000 | Rubber and plastic | 2.6 |
| 3100 | Leather | 0.17 |
| 3200 | Stone and clay | 2.4 |
| 3300 | Primary metals | 24.0 |
| 3400 | Fabricated metals | 1.7 |
| 3500 | Nonelectrical machinery | 2.6 |
| 3600 | Electrical machinery | 1.7 |
| 3700 | Transportation equipment | 1.3 |
| 3800 | Professional and scientific institutions | 0.12 |
| 3900 | Miscellaneous manufacturing | 0.14 |

Source: Weston (1970).

The choice of waste container is important because it determines the ease with which the material can be handled or disposed of. Plastic bags have come to be widely used for commercial and institutional purposes because they are lightweight and flexible. However, they tear easily, are generally not biodegradable when deposited in landfills, and create noxious fumes when incinerated.

Most suburban residential areas require that wastes be brought to the curb in galvanized or plastic 30-gallon containers, which are more desirable from an aesthetic point of view and can easily be handled by one worker. Larger plastic containers of 82-gallon capacity mounted on wheels or on special carts are also used, generally in connection with collection trucks having mechanized loading equipment (Wilson, 1977).

*Dumpsters* are wheeled metal containers with a capacity of 1, 2, or 3 cubic yards; they are regularly used on construction projects and frequently in large commercial developments, at institutions, and in some multifamily residential applications. Their contents are loaded on trucks by mechanical means.

Trucks used for solid waste collection come in a variety of sizes and shapes. Packer trucks range from 15- to 33-cubic yard capacity; these are loaded either on one side or at the rear, and the material is mechanically compacted within the truck body to optimize its carting capability. In suburban areas, particularly where the topography is hilly, smaller vehicles such as three-wheeled scooters or half-ton open-bed trucks are used as satellites to the larger packer truck. A standard truck must always be available for bulky items (mattresses, refrigerators) and for yard wastes.

As labor costs are the major part of every solid waste collection budget, truck design can be of crucial importance to the process. Although the size of collection crews varies from one to three workers, some vehicles are designed to operate with one-person crews when using mechanical loading devices. Innovation in this area continues.

Since landfills, which are usually the ultimate destination, are generally located at a distance from the urbanized areas where the wastes are collected, it is frequently desirable to provide transfer stations. The packer trucks are discharged at these locations and the wastes reloaded into 40-cubic-yard containers for the final long haul to the landfill. This saves time and permits the dump trucks to resume their collection routes more quickly than if they had to go to the landfill themselves. Transfer stations are always justified when a trip to the landfill exceeds 60 minutes.

The ultimate collection method is the vacuum system. Chutes from high-rise buildings or ground-level stations are connected to an underground pipe loop which is maintained under a high vacuum. Automatically controlled gates in the piped connection between the chutes and the loop are opened in sequence intermittently, allowing wastes to be sucked to a central point as far as a mile distant. The central point can be a transfer station where the wastes are compacted into containers and hauled away; or the wastes can be brought to the charging room of an incinerator to be processed directly. The first costs of such a system are extremely high: the 24-inch-diameter steel pipe must have very heavy walls to withstand continuous abrasion; cathodic protection is essential to avoid corrosion; change in the direction of the pipe must be designed to assure smooth flow, which requires a great deal of space; and the control system must be highly reliable. The energy cost of maintaining the vacuum at all times is also considerable. Nevertheless, such systems have been applied successfully at Roosevelt Island in New York City and at the Disney amusement parks. In hospitals vacuum systems are used to move laundry as well as solid wastes (Gibbs and Hill, 1971).

## Treatment Methods

In order to determine how to treat the solid waste stream, it is important to understand its composition. A typical sample of New York City waste is composed of the following materials by percent of weight (NYC Department of Sanitation, 1982):

| Burnables | | Nonburnables | |
|---|---|---|---|
| Food wastes | 18.1% | Glass and ceramics | 8.1% |
| Paper products | 52.5% | Metals | 7.5% |
| Textiles and plastics | 3.9% | Brick, rock, dirt | 2.6% |
| Leather and rubber | 0.6% | Miscellaneous | 1.0% |
| Yard wastes | 4.9% | | 19.2% |
| Wood | 0.8% | | |
| | 80.8% | | |

The most desirable method of solid waste management is the separation of waste streams at the source. If most of the paper products were collected separately at the source, 50% of the management problem would be removed. At least another 10% could be reduced through the recycling of glass and cans. If source separation could be mandated at the local level (and many municipalities are beginning to do so), and with the cooperation of industry in utilizing recycled materials, the solid waste management issue could be drastically reduced. However, in the United States, raw materials are still too inexpensive to make recycling generally acceptable. We recycle 25% of our paper and 7% of our glass. Japan recycles 50% of its paper and 43% of its glass, because its natural resources (as well as its landfill areas) are more limited (Herschkowitz, 1987).

Many buildings practice *compaction*. Wastes come down the chutes of high-rise structures directly into the hoppers of compactors. The presence of wastes triggers the operation of a ram, which compresses the wastes into paper containers. When one container is full, a turntable brings the next container into line for filling. The resulting package is easily handled for carting away. Compaction densifies the wastes from 200 to 600 pounds per cubic yard. Rodent and insect sprays are injected into the containers during compaction. Energy costs for the compactor are generally insignificant.

Office buildings and stores discard large quantities of paper and cartons. Baling machines compress these materials into readily handled packages held together with wire. Large food-processing operations occasionally resort to *pulping*. In this process water is added to emulsify the wastes, which are then compressed into bundles and dried for carting. In addition to requiring large quantities of water, the process is very energy intensive. *Air classification* is practiced in connection with several disposal processes. Wastes are blown through a chamber at high velocity. The heavy fractions, such as glass, masonry, and metals, settle out and allow the lightweight balance of the material to be collected for incineration.

*Composting* is the aerobic decomposition of organic wastes, such as leaves, waste paper, food wastes, grass, and sewage sludge, to produce a humuslike material. Microorganisms fueled by oxygen (similar to a sewage treatment plant) cause the organic material to heat and decay into a useful fertilizer and soil conditioner. Although many attempts have been made to start composting operations on a community-wide scale, it has generally been difficult to get people to use the resulting material in the quantities in which it is being produced (Epstein and Williams, 1988).

## Disposal Methods

In the past open dumps in isolated areas of a town or county were considered adequate for solid waste disposal. As our awareness of environmental problems became more sophisticated it was realized that in addition to odors, dumps contributed to water pollution and bred rodents. To overcome these shortcomings, landfill procedures were developed.

A state-of-the-art landfill is made up of cells; these are confined areas in which solid waste is spread and compacted in layers by bulldozers. At the end of each day, or more frequently, the waste is covered with a thin layer of soil, which is also compacted. Decomposition of organic materials is hastened in that way. When a cell becomes full, the operation is repeated in the next cell.

In order to contain leachate, landfills must be lined with synthetic material or impermeable clay soils. The leachate is collected and, if necessary, treated before disposal. Each landfill must have at least three groundwater monitoring wells where samples are collected regularly and tested for possible escaping leachate (Robinson, 1986). The decomposition process also generates methane gas. Perforated pipes are buried throughout the landfill to collect this gas for heating use, or to vent it to the atmosphere to avoid inadvertent explosions.

At present, 90% of unprocessed waste in the United States is landfilled. It is evident that suitable sites within reasonable distance from urbanized areas are becoming more difficult to find. It must be remembered that landfill will always be required not only because certain material cannot be recovered or processed, but also because there is always residue from incineration processes.

Incineration reduces solid wastes 90% by volume and 75% by weight, which can extend the potential life of existing landfills by many years. While land pollution can be mitigated in this way, air pollution may be created instead. Incinerators can be provided with electrostatic precipitators or *baghouses* (filters) to remove particulates in their effluents, or they can be fitted with *scrubbers*, which also remove some of the gases from their discharges.

In small-scale incinerator installations such as apartment houses, the first cost of such pollution control is very high and the operation of the devices is generally not well supervised. This has led to the abandonment of such on-site incinerators in favor of municipal operations, where pollution control is more cost-effective and more efficient. Central incineration has the added advantage of making heat recovery possible. An average pound of solid waste contains 5000 Btu of thermal energy, which can be used for heating, process use, electric power production, or even cooling. The use of heat recovery incinerators, or waste-to-energy plants (WTEs), is becoming more widely accepted.

Early examples of this disposal method are the Nashville Thermal Energy Company, which accepts solid wastes for a tipping fee and sells high-temperature water and chilled water to office buildings in the central business district (Anon., 1987) and the Saugus, Massachusetts, plant, which accepts solid wastes from 14 communities and produces process steam for a neighboring industrial plant. These operations suffered from early technical problems, including inadequate air pollution control, but both are now working well. A more recent project is the Peekskill, New York, plant built by Westchester County to accept solid wastes from its communities and to produce electricity, which is sold to a local utility. At the end of 1987, 200 WTE plants were either completed or in various stages of planning (Gershman and Petersen, 1988). Given the constraints on land available for landfill, it is inevitable that waste-to-energy plants will proliferate. New techniques will be developed to reduce air pollution and to reduce costs. Fluidized-bed combustion (see Chapter 4) and pyrolysis, or starved air combustion, will become more prevalent.

Since the quality and composition of solid wastes as received at the incinerator can vary from season to season or even from day to day, operators would prefer to preprocess the waste stream. Many processes to develop refuse-derived fuels (RDFs), such as grinding, cutting, sintering, or the making of pellets, have been explored but have generally been found to be too expensive or too cumbersome. Thus mass burning after rudimentary source separation prevails. Ocean disposal was practiced by many cities located near open waters. Except for isolated instances where hardship has been claimed in the disposal of sludge and construction wastes, ocean dumping has been prohibited by the EPA and the remaining locations are being phased out rapidly.

## Planning Considerations

One of the critical considerations of planning for effective solid waste management is the composition of our waste streams. At present the quantities of plastics being discarded is so large that their disposal creates serious problems. In landfills, these substances, which are not biodegradable, inhibit the decomposition of organic wastes. When burned in incinerators, plastics foul grates, requiring shutdowns and extra maintenance. Combustion of plastic substances creates noxious fumes which must be scrubbed out of the effluent before being discharged to the atmosphere. One solution would be prohibition of the use of plastic materials; this may seem farfetched, but it is being practiced by certain jurisdictions, such as the state of Vermont and the city of New York, which no longer permit the purchase of plastic eating utensils for their in-house eating facilities. When the scarcity of landfill and the exorbitant cost of disposal become more widely understood, such radical solutions will become more palatable.

When designing residential subdivisions it is important to plan for waste collection, which can be done from the backyard or the front yard; the location of service and access roads is determined by such decisions, as is the cost of collection. Road patterns for each new area must be planned to accommodate packer trucks and other collection vehicles. Road widths have to permit the passage of a car while waste collection is going on; however, excess widths encourage on-street parking, which in turn makes waste collection more difficult. Climate affects the frequency of collection: Whereas once or twice weekly may be acceptable in cooler northern climates, a minimum of three times per week will be required in warm climates to prevent the propagation of odors.

The location of landfills or incinerators is a major planning problem. Landfills require large open areas, preferably shielded from view; they must be far enough from inhabited areas to make their presence acceptable, yet they must be close enough to avoid lengthy round trips for the collection vehicles, and roads leading to them must have the capacity to accept the intensive traffic. As urbanized areas are spreading and environmental regulations for landfills are becoming more restrictive, incineration is becoming more attractive as a disposal method. Incinerators also generate traffic; they also tend to create air pollution, which must be controlled at considerable expense. Rodent control is required for the storage areas that contain the wastes prior to firing.

Solid waste collection is performed by both public and private entities. In New York City, for example, residential and institutional wastes are hauled by the city department of sanitation, while commercial and industrial wastes are collected by private carters. Privatization has become the watchword of public finance in recent years, and from the municipality's point of view it makes sense to get competitive bids and to let the successful contractor worry about staffing and fringe benefits. However, in smaller localities solid waste collection departments provide patronage jobs and employment opportunities for those who are difficult to employ.

## Costs

Costs for the construction of waste-to-energy plants burning 750 to 2000 tons per day are reported to range from $84,000 to $145,000 per ton per day of capacity, including construction, financing, and insurance. There does not appear to be a cost advantage due to scale alone; however, the lowest-cost plant burned refuse-derived fuel, whereas all the others were mass-burning plants (Anon., 1984). Nationally, the average cost of solid waste management is $25 per ton, which includes $18 for collection and $7 for disposal. This cost reflects landfilling practice. New York City costs, for comparison, are $45 per ton. Where incineration is practiced, costs are generally high, because tipping fees (charges for permission to dump the wastes) range from $10 to $74 per ton. With environmental restrictions on the increase, and landfill locations becoming scarce, it is estimated that the cost of solid waste disposal will increase to $112 to $138 per ton (Gershman and Petersen, 1988).

# HAZARDOUS WASTES

Hazardous waste can be in the form of solids, liquids, or gases. Inasmuch as these wastes are proliferating, they are becoming more important not only as an issue for treatment and disposal but also as an economic problem.

## Types of Hazardous Waste

Hazardous wastes include the following:

Flammable, explosive, and radioactive materials
Materials that are toxic to human beings and/or the environment, including plants and animals

It has been estimated that 50,000 new chemicals have been developed since World War II, and new ones are being formulated at the rate of 1000 per year. It is probable that 7 out of 10 of these materials are harmful to the ecology. Yet only 717 substances have been identified and listed as requiring federal regulation as of 1987 (Government Institutes, Inc., 1988). Industry is the primary source of these wastes; in 1986 alone it was responsible for 260 million metric tons of known generation. There are an estimated 650,000 generators, but only 2% of these are responsible for 99.6% of the wastes generated. The largest fraction came from industrial organic chemicals (18.5%), followed by industrial inorganic chemicals (16.2%), and petroleum refining (10.4%) (Apogee Research, Inc., 1987).

Commercial sources include gasoline stations, dry cleaning establishments, pesticides, and household aerosols. Hospitals are a source of hazardous wastes, with quantities continually on the increase; infectious wastes are generally required to be placed in red bags clearly identified as to source. Radioactive wastes from experimentation and nuclear power plants are a long-term problem in search of a solution under the control of the Atomic Energy Commission.

## Treatment and Disposal Methods

Thermal destruction technologies include all types of incineration, including liquid injection, rotary kilns, fluidized beds, and industrial boilers. The method used depends on whether the wastes are liquid, sludge, or solid, and on the heating value of the material. As in the case of municipal solid wastes, incineration is used either just to burn the material, or for the purpose of utilizing the heat of combustion. The latter method is gaining in application, both for environmental and economic reasons and because of shortages in qualified disposal sites.

Hazardous waste treatment technologies include a wide variety of chemical, biological, and physical processes that remove hazardous constituents from wastewaters or convert hazardous waste into less hazardous forms. These technologies are carried out either at the industrial site where they are generated, or off-site by commercial service companies. A wide variety of innovative treatment methods are being tested all the time; their success will depend on the amounts of time and money invested in them (Apogee Research, Inc., 1987).

The regulations for land disposal of hazardous waste are becoming more restrictive under both federal and state regulation. Double lining of landfill sites is being required to avoid any possible leaching of hazardous material into the ground. *Deep well injection* is widely practiced. This method involves the pumping of hazardous wastes into saline, permeable injection zones thousands of feet below the surface, well below drinking water aquifers. Federal permits are required for drilling these wells; although no known problems exist with drinking water pollution from these injection wells, there is no way of determining the effect of the injected materials on the rock layers that contain them.

Burial of radioactive wastes in unused salt mines is being tried to establish whether these caves will contain such materials over the long run. Glass encapsulation of containers holding radioactive wastes is a method that holds promise of success, although at considerable cost. Ocean incineration of liquid hazardous wastes is being practiced in Europe. Resistance to this method in the United States has centered around the dangers of spills during handling of these materials, as well as the creation of air pollution as a possible by-product of the products of combustion.

## Legislation

A Solid Waste Disposal Act was passed in 1965. In 1970 the Resource Recovery Act passed as an amendment to the 1965 law. The latter provided technical and financial assistance to states in developing resource recovery and other disposal technologies. The Resource Conservation and Recovery Act (P.L. 94-580) was enacted in 1976 as an amendment to the 1970 law to deal with the management of hazardous wastes. Not until 1980 did the EPA issue the first standards in connection with the generation and transportation of hazardous wastes. Dissatisfied with this performance, Congress set a firm timetable for the implementation of specific rules when RCRA was reauthorized in 1984. This included the regulation of underground storage tanks, the promulgation of rules for small-quantity generators, and the total prohibition of land disposal for certain wastes by specific dates (Government Institutes, Inc., 1988).

Various levels of enforcement authority have been delegated to the states. Compliance has generally not been good and as rules become stricter, more waste generators fail to be in compliance. Federal funding and grants to the states for this purpose have been increasing.

In 1980 Congress passed the Comprehensive Environmental Response Compensation and Liability Act (CERCLA), also known as the Superfund Program. It required EPA to compile a National Priorities list (NPL) of 400 sites that would qualify for federally administered cleanup. This was paid for by the $1.6 billion superfund, financed from taxes on chemical and petroleum feedstocks and some federal appropriations. At the end of 6 years the NPL had grown to 900 sites, with the possibility of 100,000 more being identified in the future. The superfund was expanded to $8.5 billion, to be financed out of increased taxes on the feedstocks as well as new taxes on domestic and imported oil (Apogee Research, Inc., 1987).

A new Title III established detailed reporting obligations for companies using hazardous substances which identifies their path from "cradle to grave" and will facilitate their management and control. Many states have passed superfund legislation of their own to finance their share of the federal cleanup effort.

## Planning Considerations

The siting of hazardous waste disposal facilities is plagued by the NIMBY ("not in my backyard") syndrome. Most states have conducted surveys of hazardous waste generated within their borders and now have adequate information on the magnitude of the problem. Areas will have to be identified, hearings will have to be held, and locations will have to be chosen. Local land-use planners will be deeply involved with the process, because their awareness of the link between economic activity and waste handling needs will be crucial to the success of the site development process.

One of the most effective hazardous waste management methods is the reduction of waste generation. Some of the more effective waste reduction strategies are as follows (Apogee Research, Inc., 1987):

Waste segregation, to avoid mixing small quantities of hazardous wastes with large quantities of normal wastes

Improvements in plant operations, such as housekeeping, materials handling, and process monitoring, to reduce the generation of hazardous wastes

Raw material modifications to introduce fewer hazardous substances into the production process

Recovery and recycling of hazardous waste constituents at the point of generation

End-product substitution of products that require fewer hazardous constituents in the first place

Information and education will play a major role in achieving these goals.

## Costs

The cost of disposing of hazardous waste is very high. Where a generator must send its wastes to an off-site landfill that can accept the material, it may have to pay $240 per ton. If the material must be incinerated so as not to create air, land, or water pollution, the cost may run as high as $800 per ton. As landfill sites in the vicinity of urban areas become more difficult to find, hazardous wastes are being shipped farther afield, incurring even higher cost penalties. Ultimately, these costs will spawn new technology and will force generators to practice waste reduction.

## CONCLUSION

Solid and hazardous waste infrastructure does not suffer so much from disrepair as in the case of water mains or highways; however, there is clearly a shortage of capacity for dealing with the waste streams, technology is not being applied as effectively as it could be, and governmental enforcement is constrained by the lack of adequate funding. To correct these deficiencies is the task for professionals in the years ahead.

# REFERENCES

Anon., Resource Recovery Rating Approach, *S & P Credit Week*, June 4, 1984.

Anon., Nashville Facility Heats Downtown Office Buildings, *American City and County*, August 1987.

Apogee Research, Inc., *Hazardous Waste Management*, National Council on Public Works Improvement, Washington, D.C., 1987.

R. W. Beck & Associates, *Solid Wastes*, National Council on PUblic Works Improvement, Washington, D.C., 1987.

Epstein, E., and T. Williams, Solid Waste Composting Gains Credence, *Solid Waste and Power*, June 1988.

Flack & Kurtz Consulting Engineers, *Solid Waste Management and Energy Supply*, Harbison Development Corporation, Harbison, S.C., 1974.

Gershman, W. G., and N. M. Petersen, WTE Development: A Zig-Zag Course, *Solid Waste and Power*, August 1988.

Gibbs and Hill, *Research Study on Refuse Collection and Disposal for Welfare Island*, Welfare Island Development Corp., New York, 1971.

Government Institutes, Inc., *Environmental Laws and Regulations: Compliance Course*, GII, Rockville, Md., 1988.

Herschkowitz, Allen, *Garbage Management in Japan: Leading the Way*, INFORM, New York, 1987.

NYC Department of Sanitation, *Resource Recovery and Waste Disposal Planning*, New York, 1982.

Robinson, William D., *The Solid Waste Handbook: A Practical Guide*, Wiley, New York, 1986.

Weston, Roy F., *Industrial Waste Generation*, Weston, Co., West Chester, Pa., 1970.

Wilson, David G., ed. *Handbook of Solid Waste Management*, Van Nostrand Reinhold, New York, 1977.

# ADDITIONAL READING

Lehman, J. P., ed., *Hazardous Waste Disposal*, U.S. EPA, Washington, D.C., 1983.

Niessen, Walter R., *Combustion and Incineration Processes*, Marcel Dekker, New York, 1978.

# QUESTIONS

1. How many pounds of solid waste per day is generated by the average person?

2. What are the most common types of equipment used in solid waste collection? Give a range of sizes for each.

3. Describe the most common disposal method in the United States. What environmental precautions must be taken?

4. Describe at least two planning considerations involved in solid waste management.

5. How is the by-product of solid waste incineration utilized?

6. Describe the types and sources of solid wastes.

7. What is the preferred treatment of hazardous waste?

8. Name the federal legislation that provides funds for the cleanup of hazardous wastes.

9. Name two methods of hazardous waste management to be emphasized in the future.

10. List the broad categories of solid waste generation.

Con Edison's Ravenswood Power Plant

# 4

# ENERGY

**George Rainer**
Flack & Kurtz Consulting Engineers

# INTRODUCTION

The supply of energy to urban areas is closely linked to their economic and social development. The advent of the modern steam engine in 1769 revolutionized transportation modes and fueled the industrial revolution. When the transmission of electric power became widespread at the beginning of the twentieth century it gave a further boost to industrial production and started the introduction of amenities into the everyday lives of Americans.

Until 1974 the growth of gross national product (GNP) in the United States was closely related to the increase in energy consumption. When the oil embargo of that year made our dependence on foreign sources more obvious, a newly found conservation ethic was able to reduce our consumption patterns in a major way, curtailing energy consumption without restricting growth (Forscher, 1989).

The United States, together with most other industrialized nations, is still dependent on offshore sources of energy. Due to the finite supply of oil and gas in the world, together with the potential of political upheaval in the Middle East, it is important that design professionals should understand the implications of the energy supply and consumption needs caused by the developments they produce, and to recognize the methods by which this impact can best be mitigated.

# ENERGY SOURCES

The forms of energy that are of primary concern to urban infrastructure are thermal and electrical. Inasmuch as their generation and utilization are closely related, they must be considered together. Some basic rules are as follows:

Energy cannot be created; it can only be converted from one form to another.

Energy always flows from a higher level to a lower level.

Thermal energy is generated through the combustion of fuels, through nuclear fission, or it can be collected directly from the sun.

## Fossil Fuels

Fossil fuels, such as oil, gas, or coal, are burned in the combustion chambers of boilers, which are large vessels containing water. The heated water or steam is circulated through pipes to process or comfort heating equipment, where it gives up its heat content and is returned to the boiler for reheating.

Coal is the most readily available fossil fuel. The United States has known deposits that could satisfy the nation's needs at current consumption levels for hundreds of years in the future. However, there are many problems. Underground mining, as currently practiced, is not safe. Strip mining plays havoc with the landscape wherever it occurs. The ash from certain types of coals contains metals, which create problems with its disposal. The products of coal combustion leaving the boiler plant chimney contain soot, as well as oxides of sulfur and nitrogen, which are injurious to health. Although the particulates can be removed from the effluent by means of electrostatic precipitation (at a considerable cost), removal of the oxides is more difficult. Effluent is passed through scrubbers using calcium carbonate reagents, resulting in liquid wastes that require special care in their disposal.

Some progress has been made in minimizing some of these problems. An alternative method of burning coal is the *fluidized-bed* combustion process, where coal is burned in a bed of air-suspended particles of limestone or sand. Emissions from this process meet EPA regulations for limits on sulfur oxides without additional treatment. Due to relatively low operating temperatures, nitrogen oxide emissions are inherently low and no additional scrubbing is required. Particulate emissions can be controlled with baghouse filters. Many installations of this more efficient process exist, mostly in industrial applications, but commercial and residential applications are limited and conversion of existing boilers is practically impossible.

Oil is injected into the boiler's combustion chamber under pressure. U.S. Commercial Grade No. 6 oil requires preheating before it will burn; USCG No. 2 oil burns more readily but is more expensive. To minimize the emission of sulfur oxides resulting from the combustion of oil, the percentage of sulfur content permissible is regulated by law.

Natural gas is the cleanest fuel to burn. Its availability is limited to areas where pipelines exist. In smaller quantities liquefied gas can be made available in refillable bottles and tanks, but this method becomes impractical with larger installations.

Nuclear fission is the controlled atomic explosion of uranium. The process is carried out in a submerged vessel, and the heat generated turns water into steam, which can then be utilized in the conventional manner. Accidental radiation spills from such installations can play havoc with the environment, but when properly operated, nuclear energy is pollution free.

## Solar Energy

Solar energy can be collected actively by means of various conversion technologies:

Flat-plate collectors
Concentrating collectors
Photovoltaic cells
Power towers

The technology that has reached the highest level of commercialization is the *flat-plate collector,* which is available from several manufacturers. Flat-plate collectors are best applied in a dispersed manner (i.e., on a building-by-building basis) such as for domestic hot water heating, which is a year-round requirement, thus making the application more cost-effective. Economic feasibility can generally not be demonstrated. One square foot of an average flat-plate collector in an average U.S. location collects the energy equivalent of 2 gallons of No. 2 fuel oil per year, which at a price of $1.00 per gallon saves $2.00 per year. Each square foot of collector installed, including piping, insulation, pumps, tank, and controls costs a minimum of $40. This indicates a payback period of about 20 years, not a good economic choice.

*Concentrating collectors* use parabolic troughs or evacuated glass tubes and produce higher-temperature liquids at a much higher cost. The higher temperatures are generally not needed for conventional heating systems. Although absorption refrigeration machines could be powered by these higher-temperature fluids, the cost—above $5000 per ton of refrigeration (compared to $1000 for more conventional methods)—has proven to be prohibitive.

*Photovoltaic cells,* which convert sunlight directly into electricity, have a favorable potential. Extensive research is being carried out to reduce the cost of these cells, which were developed to power the instrumentation of spacecraft. Costs still run between $7000 and $10,000 per kilowatt, compared to approximately $1000 to $1500 per kilowatt for fossil-fueled plants. Nevertheless, applications in areas remote from other power supplies have been shown to be practical. In the megawatt range, land areas of 10 to 20 acres are required (Clutter, 1987).

*Power towers* are pressure vessels (boilers) installed at the top of elevated structures, which are irradiated by large fields of mirrors reflecting the sun. The direction of the mirrors is computer controlled so that they are perpendicular to the sun at all times. Steam is produced at high pressures; this can be converted to electricity by means of conventional steam turbines. Large open areas of land (such as deserts) are required for this purpose. The best known installation of this type was built in Barstow, California, by Pacific Gas and Electric Co. in cooperation with the U.S. Department of Energy, and has proven to be successful.

## Alternative Sources

Alternative sources of energy include wind power, biomass, solid wastes, and geothermal energy. *Wind power* technology has been available for centuries, but the size of these windmills was generally small. In recent years the U.S. Department of Energy has embarked on a program of proving the feasibility of machines in the range of 2 to 5 megawatts. Several different types were developed, and these are now operating as wind farms in the east–west valleys of California, where winds of adequate velocity are prevalent. As of 1986, 15,451 turbines having a capacity of 1267 megawatts had been installed. To be effective, winds of 10 mph or more must be available during a sufficient number of hours per year to make the use of windmills economically feasible. Wind data for a number of years is generally available from the National Oceanic and Atmospheric Administration (NOAA) for a large number of locations throughout the United States (NOAA, 1987). Technical considerations in applying wind power are the necessity for backup power when wind is not available, as well as the interconnection of wind-generated power with conventional power grids, which can be costly. Environmental concerns may be caused by the noise generated by windmill blades, but since these installations are generally remote from settled areas, the problem may not arise.

*Biomass* technology derives energy from the burning of shrubs, grasses, crop wastes, or trees. The problems inherent with this technology are similar to those found with solid wastes combustion: transportation, storage, pollutant emissions, and residue management. Such processes would have to be carried out in a central plant, preferably with the use of fluidized-bed boilers, described above. Other known biomass processes include low-Btu gasification, fermentation, anaerobic digestion, and similar thermochemical conversion methods. As described previously, the waste gas produced by a sewage treatment plant can readily be used for heating or other power purposes (Halacy, 1987).

The *combustion of solid wastes* is a natural source of thermal energy. As we have learned, solid wastes can be disposed of in landfills or through incineration. Since open space suitable for disposal located in the vicinity of urban areas is dwindling rapidly, reduction through burning is becoming more widely practiced. It is interesting to note that the solid wastes generated from a large housing development generally contain sufficient thermal energy to provide the domestic hot water requirements for the people living there. Since both the solid waste stream and the hot water requirements are a year-round occurrence, they provide the possibility for successful application of such an installation.

Solid wastes can be burned directly in heat recovery incinerators with supplementary burners in the effluent to complete the combustion process. A more sophisticated approach is *pyrolysis, starved air incineration.* This process can generate heat and/or produce fuel oil or gas. Solid wastes can also be processed into refuse-derived fuel (RFD). After adequate separation into different waste fractions (glass, metal, masonry, combustibles), the combustible portion is processed into a high-heat-content substance, which can be burned in combination with coal. Although the cost of RFD is close to that of coal, credit can be taken for the recyclable materials taken out, as well as for the reduced carting costs of the refuse.

*Geothermal energy* is derived from the heat content of the lower strata of the earth. The most obvious manifestation of this heat can be found in geysers such as those in Yellowstone Park. Some of these warm springs have been harnessed to provide heating to settled areas; the most famous incidence is the town of Rekiavik in Iceland, where the hot water from geysers a few miles outside has been piped to the city and heats the community in its entirety. However, geothermal energy can be found in many places by drilling holes many miles deep into the earth. The hot fluids that exist under pressure in the lower layers of the earth's crust, usually in the form of brines, can be brought to the surface and are used either to drive turbines to generate electricity, or are used for thermal heating directly. Geothermal fluids generally contain high levels of dissolved minerals and salts and are therefore highly corrosive. The equipment in contact with the fluids must either be constructed to withstand the corrosion, at a cost, or the fluids must be treated before use. Disposal of untreated fluids can be difficult, for environmental reasons; one solution is the drilling of disposal wells which return the fluids to the layers where they came from, which doubles the drilling costs. Treated fluids, on the other hand, can be used for irrigation and other purposes. Geothermal energy can also be tapped by drilling two holes into hot rock, then pumping water down one hole and obtaining hot water up the other. Some successful installations exist in California and Utah, but the widespread application of the method awaits the development of more economical drilling technology.

## Electric Energy Generation

A generator for producing electric power is a rotor wound with wires rotating in a magnetic field. The rotation is produced by a prime mover such as a diesel engine or a turbine which may be energized by steam, gas, or water power. The most common method in use is the steam turbine generator, which obtains its steam from a high-pressure boiler plant, fired by any of the fuels described above.

The planner's concern with power plants is their siting. They should be near a source of fuel; this could be near a natural gas pipeline, near an oil tanker terminal, adjacent to a coal mine, or at least near a railroad siding used for bringing the fuel to the plant. Although electricity can be transported over wires for considerable distances, the power plant should be in the vicinity of the power users. Water for cooling purposes must be available, so the proximity of a river or a lake is essential. In addition to the buildings housing the equipment, a power plant requires major land areas for storing fuel as well as ashes and sludge (the effluent from stack scrubbers). It is not unusual for a 500-megawatt coal-fired plant to take up 300 acres. Finally, the flue gases and their effect on the environment must be the planner's concern.

The future promises to bring more benign energy sources. The *fuel cell* has been under development for some time. Basically, this device consists of two electrodes in an electrolyte which produces electricity directly from a fuel in the presence of an oxidizer. The process represents electrolysis in reverse. Fuels used have been hydrazine and methanol, but the use of other hydrocarbons is possible. The waste product is water, and efficiencies of 50% have been experienced. Units under 5 kW have been shown to work well, but to date, problems with materials and operation have prevented successful upscaling of the fuel cell (Appleby, 1987).

Nuclear fusion is the basic theory of the breeder reactor, which would use plutonium as a fuel. Widespread experimentation is being carried out in the United States by the Department of Energy at various laboratories. The major problem appears to be the ability of materials used for containment to withstand the extreme high temperatures encountered. Also, the economic feasibility of breeder reactors in the marketplace has yet to be demonstrated. Magneto-hydrodynamics generates electricity through the interaction of hot gases with powerful magnetic fields. Both the Soviet Union and the United States have experimented in this area, but the high cost of this research has discouraged more intensive pursuit (Rosa, 1987).

## Central Plants versus Dispersed Sources

Although electric power is generally produced centrally in a power plant, the supply of heating or cooling energy and the like is usually dispersed among individual buildings. In communities of certain size and density, such as a college campus or a properly planned new town, it is frequently desirable to provide a central plant to deliver these services. Since in such a setting buildings having different functions and experience peak requirements for service at different times of day, the equipment size in a central plant is sized for only 70 to 80% of the total connected load at considerable savings in both initial and operating costs. A further advantage of a central plant is the fact that with a change in technology, conversion becomes much more practical than would be the case with dozens of dispersed plants.

It is clear that the cost of distribution piping from a central plant is higher than that for dispersed plants; however, studies have shown that at densities of 16 dwelling units per acre and above, central distribution systems can be justified, although this threshold is greatly dependent on fuel costs. In planning the infrastructure of any new community or large-scale development, the application of a central distribution system for heating and cooling must be evaluated (Conklin and Rossant, Flack & Kurtz Consulting Engineers, 1973).

### Energy Efficiency

It is essential that the efficiency of energy sources be clearly understood, not only because this determines the operating cost of such a plant, but also because it enables us to measure the improvement possible through energy conservation measures to be described. A boiler can generate thermal energy at an efficiency as high as 85%; once this boiler is combined with a system of piping and heat-using devices, the system efficiency drops to approximately 75%. The efficiency of an electric power plant is much lower; to produce 1 kilowatt (kW) of electric energy, which is equivalent to 3414 British thermal units (Btu), it requires the following thermal input (heat rate):

| | | |
|---|---|---|
| Best coal-fired plant | 8,800 Btu/kW | Eff.: 38.8% |
| Gas-turbine plant | 15,000 Btu/kW | Eff.: 22.7% |
| Con Edison Co. of N.Y. | 10,540 Btu/kW | Eff.: 32.3% |

The energy that is not utilized in these plants is lost either to the atmosphere by going up the stack as hot flue gases or by being rejected as heated cooling water. To overcome this poor thermal performance, the concept of *cogeneration* was developed; that is, electrical and thermal energy are generated simultaneously. This can be accomplished as follows: The effluent gases from a gas turbine at 1600°F piped through a heat recovery boiler will generate steam, which can be used for thermal purposes or for generating additional electricity by means of a steam turbine. By making use of the waste heat, the efficiency of the combined, or cogeneration system can be increased to 50% or more.

## ENERGY DISTRIBUTION

### Electric Power

Electric power is transmitted over wires. The longer the distance, the higher the voltage at which it is transmitted. Think of water being forced through a pipe at high pressure; the internal friction of the pipe creates a resistance to the flow and the water arrives at the other end of the pipe at a lower pressure. Similarly, the high-voltage potential at the generating end of the transmission line allows electric current to flow through the wires. The larger the cross-sectional area of the wires, the lower the resistance to the flow (and the lower the losses). Given wires of a fixed cross section, more current can be transmitted with higher voltage at the source.

At peak demand, which generally occurs on a summer afternoon, Consolidated Edison Company of New York gets help from hydroelectric sources in upstate New York or Ontario, Canada. These transmission lines operate at 360,000 volts to minimize losses over the very extensive runs. For reasons of safety this voltage is reduced in several steps until it generally enters residential buildings at 208/115 volts and commercial buildings at 480/277 volts in the New York area. These long-distance electric supplies are generally connected to a far-flung regional grid so that they can be utilized to assist a number of local utility companies.

New technology for long-distance high-voltage power lines is being developed. When placed in a cryogenic (ultralow-temperature) environment, cables almost lose their resistance, which reduces power losses dramatically. Research into superconductivity is proliferating, and this method of transmission should be available for commercial application before the end of this century.

## Natural Gas

Natural gas is distributed through pipelines which extend from the oil fields in Texas and Alaska to local utility company storage facilities throughout the United States. Availability depends on the size of the pipeline serving a given area; pressure for moving the gas over long distances is provided by means of booster compressors driven by gas-fired engines.

Some utility companies supplement their pipeline gas supplies with gas brought in barges from North Africa; others have built substitute natural gas (SNG) plants to improve their competitive advantage during energy shortages. Manufactured liquid petroleum (LP) gases are delivered in bottles or tank trucks to rural areas that lack piped supplies.

## Oil

Oil is also delivered through pipelines, which generally run from the wells to seaports, where tankers move the oil over the long haul to refineries. Local delivery is by means of barge or tank truck.

## Coal

The major method of moving coal is by means of railroad unit trains; a long string of freight cars is loaded mechanically at the mine mouth and is moved to the destination as a train, where it is again unloaded mechanically. Local delivery to smaller users is by means of trucks; or given a waterway, by means of a barge. Modern technology has made the transmission of coal slurries possible. Coal could be processed at the mine mouth and transported by means of pipelines to its destination at considerable savings. Unfortunately, it would be necessary to cross railroad rights-of-way in order to make these pipeline networks feasible. Since railroads derive major revenues from their coal shipment business, permission for these crossings has not been readily forthcoming.

## Thermal Energy

Thermal energy in the form of steam or high-temperature water can be distributed over sizable areas; such distribution through buried pipelines is practiced on college campuses, military bases, or central business districts of cities. The advantages of district heating systems were discussed earlier in connection with central plants. New York City has an extensive steam distribution system throughout lower Manhattan. Although the cost of district steam has increased from $2.50 per thousand pounds in the 1950s to $12 per thousand pounds in the 1980s, building owners still prefer this method of providing thermal energy to installing and maintaining their own boiler plant. Although few district heating systems have been built in recent years because of the high initial costs involved, there is a trend toward planning for such systems as an energy conservation measure. With this aim in mind, a new district heating system has been developed for St. Paul, Minnesota. Chilled water distribution systems over wide areas can also be very effective. Hartford, Connecticut and Nashville, Tennessee have widely used chilled water piping systems, the latter obtaining its thermal energy from the combustion of solid wastes.

## Planning Considerations

The use of utility tunnels has been conceived to avoid the need for excavation every time a buried pipe must be repaired or modified. Although their application is practiced at medical institutions and college campuses, their general use under road pavements has been found to be far too costly, even when all possible avoided costs (including social costs) have been taken into account in trying to justify them (American Public Works Association, 1971).

The distribution of energy involves several planning considerations. An attempt should be made to assign multiple uses to an utility corridor whenever possible. Telegraph lines have paralleled railroad lines for many years; fiber-optic trunks are now being installed in similar right-of-ways. In Reston, Virginia, the land under massive power lines is being used for recreational purposes, such as skiing and riding. The catenaries of power lines have a visual impact on the areas through which they run. It is essential that buffer areas be maintained between these power lines and residential uses in their vicinity. Although most larger cities run their electric power lines underground, there are still a large number of communities that transmit power by means of overhead lines, simply because it is the less costly method. In addition to being an eyesore, these overhead wires are exposed to storms and can be very dangerous in an urban setting. During the years of intensive urban renewal that followed World War II, the federal government withheld renewal funds unless power lines were placed underground. Modern subdivision regulations contain similar requirements, but the ultimate achievement of total undergrounding will not happen for many years to come.

The location of deep-water ports for the landing of fuel oil has created planning issues. Upland areas are required for transfer and storage. Means of transportation are needed for the distribution of the oil. This raises planning as well as environmental issues that must be addressed. Similarly, the location of distribution centers for the ultimate delivery of energy materials to the consumer must be carefully zoned. These centers must be close to the point of use, yet far enough from densely populated areas to avoid safety and health impacts. Finally, program budgeting for maintaining the road systems and the infrastructure are essential to keeping energy supplies moving to the consumer.

## ENERGY UTILIZATION

### Utilization Sectors

On a national basis the utilization of energy breaks down into the following approximate segments:

| | |
|---|---|
| Industrial | 37% |
| Institutional and commercial | 16% |
| Residential | 21% |
| Transportation | 26% |

The dividing lines between these uses is by no means rigid; it may be possible that offices related to industrial plants may be included under industrial. However, in broad terms these percentages are meaningful (Hirst, 1977).

### Demand versus Consumption

To assess the efficiency of energy utilization, the quantity actually used by each consumer must be measured. We are all familiar with the electric meter that is used for this purpose; the gas meter is also widely used, as is the steam meter that is used in conjunction with district heating systems. Fuel oil is measured by the gallon and coal by the ton. Each of these fuels has a known energy content which can be expressed in British thermal units (Btu). A Btu is the thermal energy required to raise 1 pound of water by 1°F from and at 60°F.

It is important to remember that there is a distinction between the energy content of a fuel at the building line where it is used, and the content at its source, where it is generated. Electric power is the most obvious example: 1 kilowatt of electricity as it is used contains 3414 Btu; where it is generated, as we have seen before, it can take from 8800 to 15,000 Btu to make 1 kilowatt. Similarly, for every pound of steam delivered to a building containing about 1000 Btu, fuel containing 1430 Btu, or more, must be burned.

Utility companies selling energy establish rate structures, which always have an energy component; that is, the consumer pays for every unit of energy used. However, there often is another component, which is known as the demand charge. This charge is levied for the highest demand in a given time frame that the customer imposes on the system. The reason for this demand charge is the fact that the utility company must maintain plant capacity and distribution networks capable of meeting this demand at all times. Units of consumption and demand for various energy forms are as follows:

| Energy Form | Demand Unit | Consumption Unit |
|---|---|---|
| Electric | kilowatt (kW) | kilowatthour (kWh) |
| Cooling | Ton of refrigeration | ton-hour |
| Heating | Btu/hr (Btuh) | Btu |
| Gas | cu ft/hr (cfh) | cu ft (cf) |

When the demand has been calculated for a given facility, it will determine the size of service or equipment that must be provided to meet the need. Consumption is related to the hours of daily, monthly, or annual use.

### Energy-Cost Equivalents

Figure 4-1 shows equivalent heating energy costs based on specified heat content of fuel and utilization efficiency. It can be used to show that electricity at 3 cents per kilowatthour costs the same as steam at $8.80 per 1000 lb, No. 2 fuel oil at $0.73 per gallon, or natural gas at $5.30 per million Btu. A similar list can be prepared for energy costs in Illinois, bringing all costs down to a basis of 1 million Btu.(*Energy User News*, Feb. 22, 1988):

| Coal | 10,752 Btu/lb | $42.54/ton | $1.98/mill. Btu |
|---|---|---|---|
| No. 2 fuel oil | 138,888 Btu/gal | $0.59/gal | $4.25/mill. Btu |

| No. 6 fuel oil | 150,000 Btu/gal | $18.27/bbl | $2.90/mill. Btu |
| Natural gas | 1000 Btu/cu ft | $3.60/1000 cu ft | $3.52/mill. Btu |
| Electric | 3414 Btu/kW | $0.119/kWhr | $34.85/mill. Btu |

The development of such a table from information available in weekly sources will be an excellent guide to the choice of fuel to be used in a given locality.

## Climatic Considerations

To plan for the energy-related requirements of a development site, a renewal area, or a new town, it is imperative that the following information be available. In the case of new development there must be a program of proposed land uses. In an existing area an inventory of existing land uses is required.

The next step is an evaluation of the weather conditions in the locality under study. Table 4-1 shows a typical climatic summary of three geographical locations being considered for the location of a new town. Heating degree days are

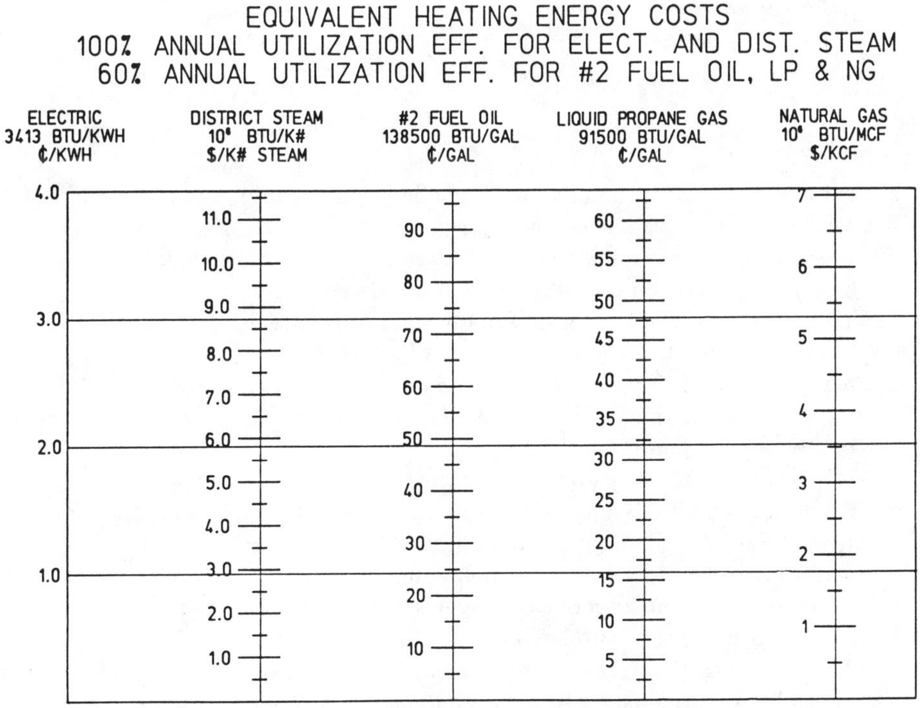

### EQUIVALENT HEATING ENERGY COSTS
### 100% ANNUAL UTILIZATION EFF. FOR ELECT. AND DIST. STEAM
### 60% ANNUAL UTILIZATION EFF. FOR #2 FUEL OIL, LP & NG

**Figure 4-1**  Equivalent heating energy cost.

**TABLE 4-1   CLIMATE SUMMARY**

| Parameter | Climate Type[a] | | | Source[b] |
|---|---|---|---|---|
| | I | II | III | |
| Heating degree days | 2601 | 6277 | 7814 | (4) |
| Winter design temperature (°F) | 28 | 1 | −4 | (1) |
| Cooling degree days | 2946 | 639 | 207 | (4) |
| Summer design temperature (°F db[c]) | 106 | 92 | 87 | (1) |
| Mean coincident (°F wb[d]) | 65 | 60 | 56 | (1) |
| db ≥ 93°F ⎫ | 943 | 79 | 2 | (3) |
| db ≥ 80°F ⎬ Hours per year | 2427 | 717 | 464 | (3) |
| wb ≥ 73°F ⎪ | 7 | 0 | 0 | (3) |
| wb ≥ 67°F ⎭ | 425 | 2 | 0 | (3) |
| Mean wind speed (mph) | 10 | 9 | 9 | (2) |
| Mean annual solar radiation (langleys/day) | 10 | 9 | 9 | (2) |
| Mean annual sky cover (in tenths, 10 = 100%) | 3 | 5 | 7 | (2) |

[a] I, Warm, dry climate; II, temperate climate: winters are cold, but not extreme; summers are cool and dry; III, winters are quite cold; summers are cool and dry.
[b] (1) ASHRAE Guide, *Fundamentals*, 1985; (2) National Oceanic and Atmospheric Administration, *Climate Atlas of the United States*, 1977; (3) *Engineering Weather Data AFM 88-29*, July 1, 1978; (4) C. Strock and R. Koral, *Handbook of Air Conditioning, Heating and Ventilation*, Industrial Press, New York.
[c] db, dry bulb.
[d] wb, wet bulb.

directly proportional to the amount of energy required for comfort heating in the location being evaluated; this is confirmed by the winter design temperature, which is the lowest for the location with the highest degree days. Cooling degree days, on the other hand, are related to the summer air conditioning requirements of the locality in question. Summer design tells you how hot it gets there, while mean coincident wet bulb is an indication of how humid the place will be at the time when summer design temperature occurs. The subsequent four lines are only a confirmation, in somewhat greater detail, of the air conditioning requirements.

Mean wind speed is of interest when wind power is to be applied. As described above, windmills must have a minimum wind speed to be effective. Mean annual solar radiation gives an indication of how effective the application of solar heating would be in the location being analyzed. The larger the available radiation, the more effective the use of solar energy. The number for mean annual sky cover merely corroborates the radiation number.

The climatic survey gives an approximate idea of what the energy requirements will be: predominantly heating, predominantly cooling, or some of each. To get a more exact reading of what the energy requirements will be, Tables 4-2

**TABLE 4-2  ENERGY DEMAND FACTORS**

| Building Use | Electric Demand (watts/sq ft) | Heating (Btuh/sq ft) for Climate Type: | | | Cooling (Sq ft/ton) for Climate Type: | | | Domestic Hot Water | Cooking |
|---|---|---|---|---|---|---|---|---|---|
| | | I | II | III | I | II | III | | |
| Residential | | | | | | | | | |
| Single family | 1.2 | 25 | 30 | 30 | 500 | 600 | 650 | 4 Btu/sq ft | 0.5 kW per dwelling unit |
| Multifamily | 1.0 | 20 | 25 | 25 | 600 | 700 | 750 | 4 Btu/sq ft | 0.5 kW/per dwelling unit |
| School | | | | | | | | | |
| K–6 | 2.7 | 25 | 30 | 30 | 300 | 400 | 500 | 500 Btu per pupil | 1 W/SF |
| 7–12 | 2.7 | 25 | 30 | 30 | 300 | 400 | 500 | 833 Btu per pupil | 1 W/SF |
| Community Facility | 5.0 | 25 | 30 | 30 | 300 | 400 | 500 | 2 Btu/sq ft | 0.25 W/SF |
| Health facility | 5.0 | 25 | 30 | 30 | 275 | 300 | 350 | 1 Btu/sq ft | 0.25 W/SF |
| Commercial | | | | | | | | | |
| Office | 3.25° | 20 | 25 | 25 | 327 | 375 | 450 | 333 Btu/person | |
| Retail | 4.0 | 15 | 20 | 20 | 300 | 350 | 400 | 2 Btu/sq ft | |
| Theater | Stage, 5.0 Hall, 1.0 | 20 | 25 | 25 | 200 | 250 | 300 | | |
| Industrial | 1 kW per employee | 25 | 30 | 30 | 300 | 400 | 500 | 1400 per employee | |
| Street lighting | 0.375 acre | | | | | | | | |

°Does not account for data processing requirements.

**TABLE 4-3  ANNUAL ENERGY CONSUMPTION FACTORS (KBtu/sq ft/yr)°**

| Building Use | Climate Type | | |
|---|---|---|---|
| | I | II | III |
| Residential | | | |
|   Single family | 30 | 35 | 40 |
|   Multifamily | 118 | 130 | 105 |
| Schools | | | |
|   K–6 | 92 | 100 | 104 |
|   7–12 | 100 | 125 | 110 |
|   College | 122 | 135 | 125 |
| Community facility | 115 | 129 | 104 |
| Health facility | 170 | 189 | 163 |
| Commercial | | | |
|   Office | 119 | 132 | 114 |
|   Retail | 150 | 168 | 144 |
|   Theater | 148 | 164 | 143 |

°Figures include energy requirements for heating, cooling, domestic hot water, fans, elevators/escalators, and lighting.

and 4-3 can be used. Table 4-2 gives a summary of typical energy demand factors. When these factors are multiplied by the square feet or the number of people called for in the program, meaningful demand indications are established. These demands can be used to size equipment (such as boilers or cooling plants) or services (such as electric power lines or gas mains). Table 4-3 gives typical energy consumption factors for a variety of uses. These factors, when multiplied by program quantities, give annual consumption (and therefore annual cost) totals for each service. In addition to being used in the engineering design of infrastructure, the demand totals, together with the consumption totals, are used extensively in analyzing the energy efficiency and the conservation potential of all types of development.

## ENERGY CONSERVATION

### Purpose

There are two basic reasons for energy conservation: (1) to conserve scarce resources, such as fossil fuels, which some day will be used up, and (2) to avoid (or delay) the need for building new power plants. Power plants represent very expensive infrastructure and must be planned for many years in advance. It is said that conservation is the most economical new capacity. Power plants built some years ago were less expensive to build (prior to inflation and at a time of fewer regulatory constraints); existing plant capacity made available for other uses through conservation efforts is therefore much cheaper than new capacity which might otherwise have to be built today. Energy conservation also saves money for facility operators by avoiding increased energy costs occasioned by higher utility rates.

## Buildings

Energy conservation in buildings starts with adequate insulation of the building shell; many guides are available for calculating an economic thickness of insulation. The use of double glazing reduces the heat transfer through windows by one-half compared to single glazing. Other methods include the building of earth shelters, where only one facade of a house built into a hillside is exposed to the weather, or underground buildings with innovative lighting from the top, where the earth acts as the insulation.

Berms can be used around buildings to deflect strong winds in cold climates, or buildings can be sited to make use of prevailing winds for natural (open-window) ventilation in more moderate climates. Waterfalls have been found to be effective in urban energy conservation. Evaporation of water requires heat, which is drawn from the surroundings of the falling water, thereby creating an inexpensive cooling effect in the vicinity of the waterfall. Vegetation can be used imaginatively around buildings to provide shading and reduce the need for air conditioning. Deciduous trees should be used on the south side of a building, so that the leaves will provide shading in summer but when the leaves are down will permit the low winter sun to help with heating in winter. Evergreens are desirable for northern exposures.

Passive solar heating has been understood for years when south-facing windows were made large to admit the winter sun. Methods have been improved by storing the captured heat in heavy masonry or in water tubes for release at night. Louvers and shading devices have been developed to prevent overheating and to retain the collected warmth indoors.

The rediscovery of daylighting has found many applications. Lighting in an office building can consume as much as 40% of the electrical energy used. The heat released from this lighting must be removed by air conditioning to make the space comfortable. With improved daylighting through light shelves at the periphery and with light wells topped by skylights in the interior of the building, you not only save the electric energy no longer needed for lighting (most of the time), but you also save the air conditioning energy no longer needed to overcome the heat of lights.

All the options described so far fall under the control of the building or the landscape architect. The building systems engineer has a broad range of options for making a facility more energy conservative. Tables 4-4 to 4-6 list energy

**TABLE 4-4   LEVEL I CHECKLIST**

| | |
|---|---|
| Operating time controls | Central start/stop |
| | Time clocks |
| | Optimal start timers |
| | Cycling timers |
| Outside air control | Reduce minimum outside air |
| | Economizer cycle |
| | Enthalpy control |
| | Warm-up/cool-down cycle |
| | Preheat coil disconnect |
| Space temperature control | Adjust set point |
| | Nonoccupancy setback |
| System temperature adjustment | Reheat system: increase fan discharge temperature |
| | Multizone or double-duct systems: raise cold deck and lower hot deck temperatures |
| | Reduce service hot water temperature |
| | Adjust heating hot water temperature |
| Lighting changes | Change bulb and ballast types |
| | Reduce number of bulbs |
| | Provide timed switches |
| | Provide additional switches |
| Escalators and elevators | Operate on demand only |
| Infiltration control | Caulk cracks at doors and windows |
| Preventive maintenance | Equipment operates more efficiently |

**TABLE 4-5   LEVEL II CHECKLIST**

Variable air volume systems
Variable volume pumping systems
Thermal storage
Free cooling
Spot cooling
Heat recovery devices
Light fixture changes
Electric distribution changes
Process changes
Building modifications
Demand controllers
Central supervisory control system
Zoning

**TABLE 4-6   LEVEL III CHECKLIST**

Equipment size reevaluation
Equipment rearrangement
Heat recovery chillers
Central plant evaluation
Cogeneration
Solid waste heat utilization
Integrated utility systems
Alternative sources of energy
    Wind power
    Hydropower
    Solar—thermal

conservation measures at three different levels, ranging from the use of simple time clocks (very inexpensive and very effective) to central supervisory control centers (featuring computerized energy management systems).

## Land Use

Energy conservation through proper land use is somewhat more difficult to achieve. The basic methods available are the creation of denser settlement, facilitation of new technology applications, and improvement of transportation systems. A study was conducted for a planned new community on the western slope of the Rocky Mountains in Colorado. Development densities were varied slightly, from 2.24 dwelling units (du) per acre for alternative 1, to 3.12 du/acre for alternative 2, to 4.4 du/acre for alternative 3. Building mechanical and electrical systems were designed for all uses, traffic model counts were undertaken, and three technology options were tested for each alternative. The results in Table 4-7 shows a meaningful range of energy reductions both with higher density and with better technology. Savings with improved technology range from 15.8 to 17.7%; savings resulting from greater density are in the same range. A maximum saving of 30.6% is shown for the use of the best technology combined with the highest density. Planned unit developments (PUDs), for example, with their higher building densities and collective open spaces, have a good potential for the application of central utility plants and innovative technology. Another more detailed example of energy savings through better technology can be seen in Figures 4-2 through 4-5 and Table 4-8, where a series of four methods of meeting the same community requirements shows that alternative energy scheme A, utilizing cogeneration, has the lowest unit energy consumption.

Transportation takes a major portion of the urban energy budget, and reduc-

**TABLE 4-7 ENERGY REQUIREMENTS FOR GRAND VALLEY NEWTOWN, COLORADO (Btu × 10⁶)**

| | Alternative 1, 2.24 du/acre | | | Alternative 2, 3.12 du/acre | | | Alternative 3, 4.4 du/acre | | |
|---|---|---|---|---|---|---|---|---|---|
| | Conventional | Central Plant | Total Energy | Conventional | Central Plant | Total Energy | Conventional | Central Plant | Total Energy |
| Energy required at user | ← 885 → | | | ← 840 → | | | ← 797 → | | |
| Raw energy required at source | 1757 | 1646 | 1425 | 1669 | 1563 | 1344 | 1579 | 1421 | 1263 |
| Transportation energy required | 350 | 350 | 350 | 288 | 288 | 288 | 200 | 200 | 200 |
| Total required | 2107 | 1996 | 1775 | 1957 | 1841 | 1632 | 1779 | 1621 | 1463 |

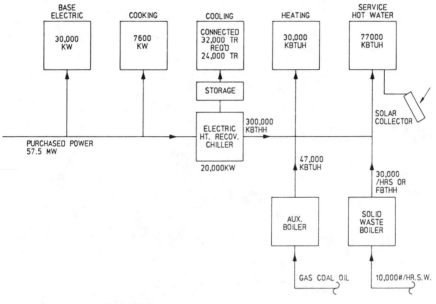

**Figure 4-2** Community I: basic energy scheme.

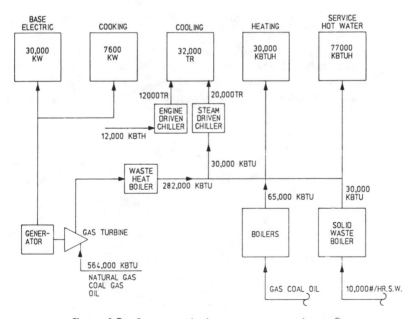

**Figure 4-3** Community I: alternative energy scheme A.

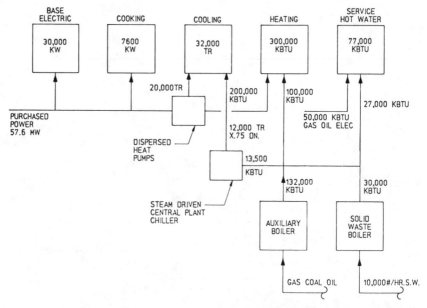

**Figure 4-4** Community I: alternative energy scheme B.

ing the need to provide it, or changing the modes that can be used, would conserve energy. Mixed-use development combines residential, commercial, and recreational functions in one area; this clearly reduces the need to travel between these different uses (Beaton et al., 1982). Mass transit can be justified only where the origins and destinations of a large number of users coincide. Building dense development at mass transit stops favors this requirement, as in

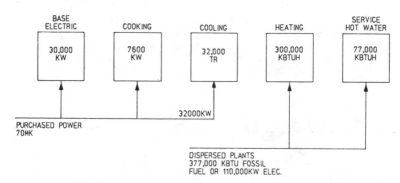

**Figure 4-5** Community I: alternative energy scheme C.

**TABLE 4-8 CONSUMPTION SUMMARY FOR COMMUNITY I**

|  | Basic | Alternative A | Alternative B | Alternative C |
|---|---|---|---|---|
| Electric demand (kW) |  |  |  |  |
| Basic | 30,000 | 30,000 | 30,000 | 30,000 |
| Cooking | 7,600 | 7,600 | 7,600 | 7,600 |
| Thermal demand (kBtuh) |  |  |  |  |
| Heating | 300,000 | 300,000 | 300,000 | 300,000 |
| Service hot water | 77,000 | 77,000 | 77,000 | 77,000 |
| Cooling demand, TR | 24,000 | 32,000 | 29,000 | 32,000 |
| Thermal (kBtuh) | — | 420,000 | 135,000 | — |
| Electric (kW) | 20,000 | — | 20,000 | 32,000 |
| Total electric (kW) |  |  |  |  |
| Summer | 57,600 | 57,600 | 57,600 | 70,000 |
| Winter | 37,600 | 37,600 | 37,600 | 37,600 |
| Total Thermal (kBtuh) |  |  |  |  |
| Summer | 77,000 | 497,000 | 212,000 | 77,000 |
| Winter | 377,000 | 377,000 | 377,000 | 377,000 |
| Source energy: |  |  |  |  |
| Fossil fuel (kBtuh) |  |  |  |  |
| Summer | — | 764,000 | 215,000 | 77,000 |
| Winter | 58,000 | 644,000 | 370,000 | 377,000 |

(continued)

**TABLE 4-8  CONSUMPTION SUMMARY FOR COMMUNITY I**

| | Basic | Alternative A | Alternative B | Alternative C |
|---|---|---|---|---|
| Purchased electric (kW) | | | | |
| Summer | 57,600 | — | 57,600 | 70,000 |
| Winter | 57,600 | — | 57,600 | — |
| Solid wastes (kBtuh) | 50,000 | 50,000 | 50,000 | 50,000 |
| Active solar (kBtuh) | 47,000 | — | — | — |
| Passive solar (kBtuh) | 150,000 | 150,000 | 150,000 | 150,000 |
| Geothermal (kBtuh) | — | — | — | — |
| Overall peak efficiency (%) | | | | |
| Summer | 41 | 76 | 44 | 36 |
| Winter | 66 | 72 | 59 | 63 |
| Annual consumption at building (Btu × 10⁶/yr) | | | | |
| Total | 1,107,848 | 1,107,848 | 1,107,848 | 1,107,848 |
| Thermal | 553,924 | 1,107,848 | 450,000 | 553,924 |
| Electrical | 553,924 | — | 657,848 | 553,924 |
| Annual consumption at source (Btu × 10⁶/yr) | | | | |
| Thermal | 692,405 | | 562,500 | 692,405 |
| Electrical | 1,846,413 | | 2,182,826 | 1,846,413 |
| Total | 2,538,818 | 2,014,269 | 2,745,326 | 2,538,818 |
| Building area (sq ft × 10⁶) | 18.9 | 18.9 | 18.9 | 18.9 |
| Unit consumption (kBtu/sq ft/year) | 134.4 | 106.5 | 145.2 | 134.4 |
| Transportation (total kBtuh/year) | 14,000 | 14,000 | 14,000 | 14,000 |
| Plant cost | Medium | High | Low | Lowest |

the case of Washington's METRO and Atlanta's MARTA systems. Building parking facilities along feeder routes can be equally effective, as in the case of San Francisco's BART system or along the bus routes leading into Denver. Despite the great cost of building subway systems, their use will have to increase, and energy conservation will only be one of the reasons.

## RULES AND REGULATIONS

The earliest effort at energy conservation standards was ASHRAE 90-75 (issued in 1975), which set numerical heat transfer standards for building components that could not be exceeded. The method has been adopted through energy-saving legislation in most states in one form or another. This standard has been updated over the years and the state codes are still being enforced, sometimes in more restrictive ways.

Another effort was started by the federal government with the help of the AIA Research Corp., which tried to develop building energy performance standards (BEPS) for typical buildings in all geographic regions of the United States. Since the energy crisis had abated before this program was completed, the funds were withdrawn and the effort abandoned; but the preliminary data gathered provide some useful guidelines (U.S. Department of Energy, 1979). At the height of the energy shortage the federal government promulgated "temperature setback" rules that were impossible to enforce and were duly abandoned.

The Power Plant Fuel Conservation Act prohibited the use of natural gas in power plants so that the limited supplies could go to residential customers or industrial processes that would have gone out of business without the fuel. However, there appear to be abundant supplies of natural gas available at this time, and that law has been repealed. Utilities within a state are under the jurisdiction of public service or utility commissions. These bodies are frequently under the influence of the companies they are supposed to control and therefore have been slow in changing wasteful practices.

Submetering of multifamily residential tenants has been found to conserve energy by as much as 17%, because making each tenant responsible for his or her own bill will limit consumption. In many areas utility companies would not permit the practice, because it meant extra meter readings for them. However, the development of central electronic meter reading equipment has overcome this objection.

Utility companies were resistant to the introduction of cogeneration because they lost business to these cogenerators who were making their own electricity; they also would not provide standby power to them except at exorbitant rates. Issues like this were addressed by PURPA, the Public Utilities Regulatory

Policy Act, which forced state utility commissions to act on these issues more favorably. After some major legal skirmishes, utility commissions are now cooperating with these conservation efforts. PURPA spawned extensive activity in cogeneration by requiring utility companies to purchase excess capacity from cogenerators at avoided cost rates; these costs vary from state to state, requiring detailed economic analysis to determine feasibility. However, the regulatory climate continues to be in a state of flux, and it is wise not to place excessive reliability on existing regulations when making economic predictions.

## REFERENCES

American Public Works Association, *Feasibility of Utility Tunnels in Urban Areas*, Special Report 39, APWA, Chicago, 1971.

Appleby, A. J., ed., *Fuel Cells: Trends in Research and Application*, Electric Power Research Institute, Palo Alto, Calif., 1987.

Beaton, W. P., J. H. Weyland, and N. Neuman, *Energy Forecasting for Planners: Transportation Models*, Rutgers University, Piscataway, N.J., 1982.

Clutter, Ted, Virginia Power Tests PV, *Solar Today*, May/June 1987.

Conklin & Rossant, Flack & Kurtz Consulting Engineers, *Grand Valley Urbanization Plan*, Report to Colony Development Corp., Atlantic Richfield Co., Denver, Colo., 1973.

Forscher, F., Energy Awareness, *Strategic Planning and Energy Management*,. Vol. 8, No. 3, 1989.

Halacy, Dan, Fuels from the Sun, *Solar Today*, May/June 1987.

Hirst, E., *Encyclopedia of Energy*, 2nd ed., McGraw-Hill, New York, 1977.

National Oceanic and Atmospheric Administration, *Selected Climatological Publications*, Environmental Information Summaries C-22, NOAA, Washington, D.C., Rev. Dec. 1987.

Rosa, Richard J., *Magnetohydrodynamics Energy Conversion*, Avco Everett Research Laboratory, Hemisphere Publishing, New York, 1987.

U.S. Department of Energy, Notice of Proposed Rulemaking, *Energy Performance Standards for New Buildings*, DOE/CS/0112, Washington, D.C., 1979.

## ADDITIONAL READING

Burchell, R. W., and David Listokin, *Energy and Land Use*, Center for Urban Policy Research, Rutgers University, Piscataway, N.J., 1982.

National Academy of Sciences, *Energy in Transition, 1985–2010*, W. H. Freeman, New York, 1979.

Real Estate Research Corporation, *The Cost of Sprawl*, prepared for the Council on Environmental Quality, U.S. Government Printing Office, Washington, D.C., 1979.

Ridgeway, James, and C. S. Projansky, *Energy Efficient Community Planning*, J. G. Press, Emmaus, Pa., 1979.

Stobaugh, Robert, and Daniel Yergin, eds., *Energy Future*, Report of the Energy Project at Harvard Business School, Random House, New York, 1979.

Touryan, K., The Wind Industry: Technology and Economic Status, *Solar Today*, May/June 1987.

## QUESTIONS

1. Describe one source each for thermal and electrical energy.

2. Describe two types of solar collectors.

3. What methods are available to obtain energy from biomass?

4. What are the advantages of using a central plant in a multibuilding development?

5. What are the advantages of cogeneration compared to providing thermal and electrical energy separately?

6. Name three fossil fuels that can be distributed through pipelines.

7. Discuss two planning concerns related to energy distribution.

8. A community in climate zone III has the following land uses:

| | |
|---|---|
| Multifamily dwelling units | 1,000,000 sq ft |
| Schools | 20,000 sq ft |
| Office buildings | 200,000 sq ft |
| Retail | 50,000 sq ft |

   (a) What is the heating demand for the community in kBtu/hr?
   (b) What is the electric demand in kilowatts?

9. What methods are available for conserving energy through proper land use?

10. Give three reasons for practicing energy conservation.

Microwave Dishes

# 5

# TELECOMMUNICATIONS

**Anthony P. Nuciforo**
Consultant

## INTRODUCTION

No one can deny the impact that the telecommunications explosion has had on our society in the past decade. The proliferation of telecommunications services and equipment has fundamentally changed the way that we conduct our professional and private lives. For the modern architect or urban planner, it is no longer enough to understand the requirements for utilities such as power, water, and HVAC services. Telecommunications is fast becoming the "fourth utility" in commercial and residential construction projects, and for this utility to be utilized properly and effectively, the architect and planner must have a good working knowledge of its requirements.

For urban areas to thrive and prosper, they must attract the large-scale commercial enterprises that will form the nucleus around which its community will grow. In today's world, telecommunications is fundamental to the success of any commercial venture. A well-planned, highly flexible telecommunications infrastructure will be the key to attracting and keeping high-tech manufacturing and service-oriented enterprises in an urban environment.

Until very recently, very little emphasis has been placed on the detailed planning of telecommunications infrastructures. While architects and planners spend a great deal of time planning roadways, airports, and recreational areas, they have traditionally left the planning of telecommunications services to other professionals, typically the local telephone company. Recent changes in the telecommunications market, and the introduction of sophisticated telecommunications services such as private fiber optic, terrestrial microwave, and satellite networks, and cable television systems, have combined to alter the traditional planning process dramatically. Because the local telephone company does not, in all probability, provide most of these services, the architect and planner cannot look to it to plan an integrated infrastructure to support them. Instead, the architect must understand the functional requirements of these various systems and be able to work with telecommunications professionals in planning an infrastructure that will support these systems today, yet be flexible enough to accommodate the systems of tomorrow.

It is not possible within the space provided in this chapter to give the architect or urban planner sufficient technical information to plan large-scale telecommunications facilities themselves. Rather, it is the intent of this chapter to provide an understanding of the basic urban telecommunications infrastructure, how it developed, and what the requirements are for the future. This information, combined with a general overview of the technologies involved and some explanation of the rather arcane technical jargon that has developed within the world of telecommunications, should allow the reader to work closely with telecommunications professionals in planning and executing this infrastructure.

# HISTORICAL OVERVIEW

Although most people do not realize it, telecommunications services have been available in the United States in one form or another for almost 150 years. In the 1840s, very simple telegraph systems were in use to provide relatively fast, cost-effective communications to what was then the western frontier. Indeed, these early telecommunications systems were also to play a big part in the growth of our industrial society, for the telegraph system was one of the first practical applications of electricity, and Thomas Edison had his first exposure to electricity when working as a telegraph operator.

Even though Edison was so impressed by the concept of communications over long distance by wire that he actually worked on developing a prototype telephone, the race was won by Alexander Graham Bell. In 1877 Bell founded the Bell Telephone Company, which began by providing rudimentary telephone services in urban areas. From this time until the middle of the twentieth century, the Bell Telephone Company, and its successor the American Telephone and Telegraph Company (AT&T), successfully staved off a wave of competitors to emerge as the single dominant supplier of telecommunications services in the United States.

By the 1920s AT&T was considered a "natural monopoly" and controlled virtually all local telephone utilities. Despite several attempts during the early and middle portions of the century to limit their control of the nation's telecommunications system, AT&T emerged in the 1950s and 1960s as the largest private monopoly in the world. At the height of its power AT&T had well over 1 million employees, $155 billion in assets, and about $90 billion in annual revenue. AT&T also controlled 80% of the local telephone service market, 98% of the long-distance market, and 95% of the telecommunications equipment market.

Despite the phenomenal growth and the success of AT&T, many problems emerged from this monopolistic approach to telecommunications service. Such a large organization, with multibillion-dollar investments in existing physical plant equipment, is naturally slow in reacting to change. From as early as the 1920s the Bell system refused to replace its existing manual switchboards with newly developed electronic switching systems. This resistance to change, and the decision to remain committed to an installed base of aging equipment, delayed the development of this automatic switching equipment in the United States and allowed several European manufacturers to gain clear advantages in this field. This resistance to technological change was to remain the hallmark of the telephone company for many years to come. Many industry analysts believe that the introduction of many of today's most common telecommunications services, from facsimile systems to satellite communications networks, was delayed for many years by AT&T's resistance to change and by the lack of competition to drive it.

## DIVESTITURE AND ITS IMPACT

For many years, AT&T resisted opening the telecommunications market to unrestrained competition. Continually stressing the benefits of a monopolistic system, and going so far as to claim that introduction of services and equipment not provided by AT&T would "harm" the public network, they successfully held off the competition. However, persistence on the part of would-be competitors gradually began to bear fruit, and by 1968 the "Carterphone Decision" granted nontelephone companies the right to provide telecommunications equipment and services to end users.

During the 1970s, the proliferation of these nontelephone or "interconnect" companies, and the wide variety of equipment and services that they could offer, created an atmosphere of total confusion within the telecommunications user community. This situation was made worse by the lack of systems standards and by the attitude of noncooperation and general hostility shown to these organizations by AT&T. This situation persisted through the 1970s and finally created an atmosphere in which the final breakup of the Bell system could occur.

In 1984, the American Telephone and Telegraph system, once the largest privately owned monopoly in the world, was dismantled by the federal government. This was the culmination of many years of antitrust litigation initiated by many of the interconnect operators. Seven regional holding companies were created to continue to provide regulated local telecommunications services. These companies were also free to sell services and equipment through unregulated subsidiaries. They retained ownership of the Bell logo as well as of the embedded local exchange equipment and cable plant. AT&T, on the other hand, was allowed to continue to provide regulated long-distance service, retained the manufacturing and R&D facilities of the old monopoly, and was freed to manufacture and market other types of services and equipment.

It was generally agreed at the time of the final antitrust decree that AT&T came out of the breakup as the clear winner. While AT&T was allowed to retain the highly lucrative long-distance services, R&D, and manufacturing facilities, the regional holding companies (RHCs) were stuck with the obsolete local physical plants and the seldom-profitable local telephone services. Despite this seemingly one-sided division of resources, it appears that divestiture has hurt AT&T and the RHCs equally. Although the federal government has allowed rate restructuring to provide the regional holding companies with added revenue to replace the lost long-distance subsidies, the RHCs have all lost substantial sums in their unregulated businesses.

AT&T has not fared much better. Having started 1984 with just over 386,000 employees, AT&T has since laid off over 100,000 and has closed many Western Electric manufacturing plants. With their market share of equipment sales down to 25% from a 1970s high of almost 100%, and a disastrous attempt at challenging industry giant IBM in the highly lucrative computer market, AT&T profits today come almost exclusively from the regulated long-distance telecommunications market. Despite inroads into this market by such competitors as MCI and Sprint, AT&T's established equipment plant and expertise in long-distance transmission technology continue to give them the edge in this market.

## REVIEW OF THE TECHNOLOGY

Before discussing any planning issues related to telecommunications systems, it will be helpful to examine briefly some of the systems and equipment that make up a typical modern metropolitan area telecommunications network.

### Central Offices

The heart of any urban telecommunications system is its network of central offices. Basic "dial tone" services, as well as the myriad of sophisticated voice and data services that are utilized in the residential and commercial environments, are all delivered to the user from the local central office. The central office is the facility that houses the large telecommunications switching and transmission systems, cable termination and distribution equipment, and all necessary supporting equipment, such as emergency batteries, generators, and air conditioning systems.

Today's telecommunications switching equipment is the result of an evolutionary process going back many decades. Originally, switching telephone calls from the point of origination to its final destination was a completely manual operation, with a long-distance call requiring dozens of separate, manual routing procedures. As telephone traffic grew, however, the processing of establishing call paths manually became both overly complicated and less cost-effective. Automatic switching systems were developed to overcome the limitations of this antiquated manual system.

Whereas the first automatic switching systems were purely mechanical in nature, newer generations of switching equipment have come to rely more and more on sophisticated electronic systems to control the routing of calls through the public network. This reliance on electronics has allowed the switching equipment to operate at steadily increasing speeds, while each successive generation of equipment grows more compact. This evolutionary process is analogous to the changes in the field of computing, where equipment has steadily grown smaller while growing in processing power and storage capacity. In fact, today's state of the art digital switching systems are almost indistinguishable from a modern computer in terms of the digital technology used in the systems architecture, the systems operation, and the facilities, such as power and cooling, that are required to support its operation.

## Private Branch Exchanges

Basically, the private branch exchange (PBX) is a miniature version of the central office switch. Typically located on the user's premises, the PBX provides basic dial-tone services, as well as more sophisticated features such as call forwarding and pickup, conference calling, and data switching. While it is true that most of these features are available directly from the telephone company's central office (this service is generally referred to as *centrex* by the local telephone companies), the biggest advantage of an on-premises PBX is the reduction in the number of central office connections, or *trunks* as they are called, that are required to service a given number of users.

Generally, if a company chooses to get telecommunications services directly from the telephone company's central office, a single circuit (usually, a single pair of wires) is required for every telephone on the customer's premises. In private residences, or small offices where few telephones are in use, this method of service poses no problems. However, in cases where hundreds or even thousands of telephones are required within a commercial or industrial complex, the quantities of cable required to provide this type of service can become quite large. Further, because these circuits are leased individually on a monthly basis, the costs for this type of service can become excessive.

The PBX helps to resolve this problem by taking advantage of the fact that not everyone in a given location will be on the telephone at the same time. Advanced traffic analysis techniques, based on average call duration and acceptable call waiting times, are utilized to determine the quantities of circuits required to service a given location adequately. On average, one circuit for every six users is a good approximation of the trunking requirements for a modern PBX. This means that an office that formerly required 1000 central office circuits to support a given number of telephone users can reduce that number to about 160 by the use of an on-premises switching system.

In addition to this obvious advantage, the PBX provides several other services that may not be available from the local telephone company. These modern PBXs can be utilized to transmit and switch data communications systems, both within a given premises and to the outside world. The PBX can also be programmed to provide detailed call accounting records and to restrict toll and long-distance calls, as well as calls to undesirable services. An added feature of an on-premises switching system is that it allows users to make their own system changes and reconfigurations without the need to call the phone company every time a phone needs to be moved. Although a PBX is an expensive piece of equipment having large power, space, and cooling requirements, the benefits that such a system provides to large-scale telecommunications users more than offsets the capital expenditure required.

## Physical Plant

The telecommunications cabling plant is the system that connects the switching equipment located at the central office to the telecommunications equipment within individual buildings. A second, physically separate cable plant is generally utilized by the utility to interconnect central office switching equipment with other local and long-distance central offices.

In a metropolitan area the telecommunications cable plant is generally located underground in duct banks, manholes, and cable vaults that are similar in size, construction, and use to the distribution system utilized by the local electrical utility. While it is true that in some areas aboveground distribution on utility poles is still in use, this method is considered undesirable, for a number of reasons. Aboveground cable distribution subjects telecommunications cables to accelerated deterioration due to prolonged exposure to inclement weather, and even newly installed cables can be completely destroyed by severe weather conditions such as tornadoes and hurricanes. Because of such factors, exposed cable plants have a much higher life-cycle cost than do underground plants of comparable size. This has caused most telephone utilities to begin replacing their outside plant cable with cable networks that are distributed underground.

Until very recently, the distribution and transmission of telecommunications signal within an urban area was done exclusively on pairs of copper wires or on coaxial cables. While the recent introduction of fiber-optic-based telecommunications systems has pushed the local telcos into installing a certain amount of fiber-optic cable in the physical plant, the vast majority of cable currently in use in both urban and rural areas is composed of paired copper conductors.

It is important to note that most of the existing cable plant in urban areas today was installed decades ago, when only minimal telecommunications requirements, and virtually no data communications, were needed. The rapid expansion of the computer into the commercial environment, and the inherent need that this equipment has to communicate, has placed a tremendous stress on this outdated cable plant. The proliferation of the high-rise office building within the past few decades has placed further stress on the cable infrastructure by increasing the population density in urban commercial centers.

Most local operating companies have initiated upgrading programs for their existing cable plant; however, this process is costly and is limited in effectiveness, for a number of reasons. The biggest problem in upgrading the urban infrastructure is the limited capacity of the physical distribution system. Many years ago, the last available space under the streets of most older urban centers was used up distributing utilities to the growing number of commercial and residential users. While the telephone company can in many cases remove older copper cables and replace them with newer ones with higher transmission capacity, the total volume of cable in a given branch of the distribution system is fixed, because new conduits cannot easily be added to existing duct banks.

These limitations in the physical distribution system have pushed telephone companies in the direction of fiber-optic cables, which utilize pulses of infrared light rather than electrical signals to transmit information. Because of their greater information-carrying capacity, a relatively small fiber-optic cable can replace many large older copper cables. While the capacity of fiber-optic cables is far greater than that of an equivalent volume of copper, fiber-optic distribution systems require sophisticated, expensive electronics to convert signals from electricity to light.

Wide-scale introduction of fiber-optic systems in urban areas has been delayed by several years, for a number of reasons. Although the cost of the cable is generally less than that of equivalent copper cable, the electronics necessary to drive the system are still quite expensive. Lack of available space in the older central offices and on user premises to house the necessary electronics, and the lack of existing industry standards for cable and transmission equipment, have all contributed to the delays in implementing large-scale fiber-optic networks.

## Bypass Technology

The term *bypass technology* is used to characterize certain types of transmission systems (such as satellites and microwave systems) that do not rely on the cabling, switching, or transmission equipment of the local telephone company to move information from one location to another. Typically, this transmission equipment is either privately owned or is leased from a independent company [known as *other common carriers* (OCCs)]. When discussing microwave systems or satellites in the context of a bypass technology, it is important not to confuse these systems with the same technologies as they are applied by the telcos for use within the public telecommunications network.

## Microwave Systems

By definition, microwave systems are radio-frequency transmission/reception systems that operate within the frequency range from 1 gigahertz (GHz; billion cycles per second) to 300 GHz. Terrestrial microwave systems that are likely to be encountered within a private bypass network typically operate at either 12, 18, or 23 GHz. At these high frequencies a signal can be subdivided into many smaller "channels" utilizing such techniques as frequency- and time-division multiplexing. These techniques allow a user to place a great deal of information, including data, video, and voice, onto a single radio channel. It is this channel capacity, and the inherent freedoms and flexibility that a private network gives, that make microwave systems attractive to many users.

Because microwave systems operate at very high frequencies, the signal broadcast by the transmitting antenna is a very narrow, highly directional beam. While this characteristic is desirable from a standpoint of limiting signal interference and preventing eavesdropping, it means that "line-of-sight" conditions must be maintained from the transmitter to the receivers. In a fully developed urban area with many high-rise office buildings, line of sight between a user's transmitter and receiver is not always available. The only solution in this case is to provide one or more "repeater" locations, to route the signal around these obstructions. This can cause the true signal path to increase greatly, adding significantly to system costs and introducing additional potential points of failure.

Despite some of these limitations, the use of microwave systems in the urban environment has grown tremendously in the past few years. In many areas, where no space is available within the telecommunications infrastructure for the installation of private cable networks, a private microwave system is a cost-effective alternative to the high cost of circuits leased from the telephone company. The largest impediment to the use of urban microwave systems in recent years has been the lack of available frequencies on which to transmit and receive.

In areas with a high density of microwave systems in use, such as New York or Chicago, interference between adjacent systems utilizing the same frequencies can cause reception and transmission problems for both. This may lead to the loss of important voice and data traffic. In fact, the data communications traffic for which most microwave systems are used is much more susceptible to interference and loss of information than is voice traffic.

The Federal Communications Commission (FCC), which regulates all radio-frequency communications within the United States, issues licenses for the use of private microwave systems. These licenses are issued for use on only certain, predefined frequencies, and users are limited to the frequencies assigned to them by the FCC. In the past several years, the FCC has literally run out of available frequencies in certain urban areas. This has caused the FCC to deny requests for service in certain frequency ranges, and the users have had to utilize more costly microwave systems which transmit at frequencies higher than those currently in widespread use.

## Satellite Systems

Like terrestrial microwave systems, satellite systems also function in the microwave frequency ranges. The basic difference between satellite and terrestrial microwave systems is the fact that where terrestrial microwave systems utilize ground-based repeaters to overcome transmission and line-of-sight limitations, the satellite systems utilize space-based repeater on satellites in a geosynchronous orbit 23,500 miles above the earth's surface.

Although satellite systems handle the same type of information as terrestrial microwave systems, their lower operating frequencies limit the amount of information that can be accommodated on a single channel. Additionally, since the signal transmitted from the satellite back to earth must travel a very great distance, it is greatly dispersed by the time it reaches the earth's surface. Although such dispersed signals are very weak and therefore require relatively large, sensitive antennas for reception, this method of transmission has the advantage of providing a "broadcast" capability to the system. This ability to cover large areas with a single transmitter allows geographically dispersed offices of a large corporation to receive, simultaneously, information sent from the home office via satellite. This is particularly useful for services such as electronic mail, daily information updates, and video and audio teleconferencing.

Because of the extremely high costs (in excess of $100 million each) involved in manufacturing satellites and placing them in geosynchronous orbits, few, if any, private companies can afford to purchase these systems for their own use. Most often, telecommunications companies such as RCA Globecomm, Western Union, and AT&T purchase and operate the satellites, leasing the on-board transmission facilities (called *transponders*) to private users. The private user need only purchase the so-called *earth station*, which is comprised of the now-familiar dish antenna and its associated transmission and reception equipment, to provide access to the satellite's transponders.

Even though this earth station equipment is orders of magnitude less costly than the satellite itself, equipment, installation, and operating costs are still quite high. Add to these costs the ongoing costs of the leased satellite circuits, and most small and midsized companies find the use of these services cost prohibitive. Additionally, many companies that could accommodate the costs of such a system are prohibited from using them because they lack the sometimes substantial amount of space that is required to house the dish antenna and its support equipment. For an antenna about 30 feet in diameter (which is typical for a system with both transmit and receive capabilities), up to 2000 square feet of space may be required for the antenna and its support equipment. Even where space is available, it may not be located within the all-important line of sight to the satellite. In urban areas, rooftop mounting of satellite systems can often overcome any line-of-sight limitations. However, large antennas have structural loading requirements (both dead weight and a sizable wind load) that may be beyond the capability of the selected structure to support.

In response to a perceived need to provide satellite services within metropolitan areas, some entrepreneurs have developed and implemented the concept of the *teleport*. A teleport is a facility that contains all the equipment that is necessary to transmit and receive signals to and from the satellites. The operators of these facilities lease large blocks of transponder time on a variety of satellites and then resell these services in packages that are cost-effective to a variety of users. Connections between this *antenna farm* and the users' locations are typically over local telecommunications circuits that are leased from the telephone company or other carriers, or they may be supported by one or more terrestrial microwave systems.

## Cellular Telephones

Several other technologies are currently in use in urban areas to provide telecommunications service. The most widely known of which is cellular telephone. This technology utilizes broadcast-type radio signals to provide a connection path from the centralized switching facilities to the individual user, which is often a mobile facility such as an automobile.

The term *cellular* is derived from the fact that the reception area of a particular system is broken up into small units called cells. Within each cell is a completely functional transmission facility, complete with transmitter, receiver, and antennas. As the user of the terminal equipment moves through a given area, the equipment will pass into and out of a number of local cells. At the central switching facility, computers that control the systems can sense the direction of travel and switch the signal from one cellular transmitter to the transmitter of the next cell.

There are several advantages to this approach to mobile communications. With standard radios, transmitters must be very powerful so that the signal that they broadcast can reach the system's single centralized antenna location. Further, within highly developed urban areas traditional radio systems are susceptible to dead spots caused by the blockage of the radio signal by large steel structures. Additionally, standard radios require individual discrete transmit and receive frequencies to prevent signal interference.

Cellular radio overcomes these limitations in a number of ways. Because antennas are located in many places, the distance between any mobile unit and an antenna is relatively small. This means that the transmitters of both fixed and mobile units can be of a lower power and thus require a smaller power supply. Further, the use of multiple antennas helps reduce the instances of signal loss due to structural interference. Because cellular transmitters have a very limited range, broadcasts from an individual transmitter can only be received within a single cell. This means that specific frequencies can be reused from cell to cell, greatly increasing the number of channels available for communications. In cases where a user transmitting on a specific frequency moves into a cell where that frequency is already in use, the central controller will automatically switch both the transmitter and receiver to a different, available frequency.

## Telecommunication Services

Much of the telecommunications infrastructure that is currently in place in metropolitan area was put in place many years ago, to provide basic voice services only. Even as recently as 15 years ago, it was not envisioned that the proliferation of computer and telecommunications systems in commercial environments would place such a heavy load on the public telecommunications network. Until very recently, the development of digital voice and data transmission systems has been a compromise between the needs of end users and the limitation of the existing telecommunications infrastructure.

## Basic Voice Services

Basic telephone service, euphemistically known within the telecommunications field as "POTS" (plain old telephone service) is typically provided to a subscriber on a simple pair of copper wires. This signal is *analog*, which means that it consists of a continuous electrical signal varying in frequency or amplitude in response to a changing physical quantity, such as the loudness of the human voice. Typically, the signal is within the frequency range 300 to 3000 hertz, which is generally considered the minimum *bandwidth* necessary to reproduce the human voice intelligibly. Because the telecommunications infrastructure was originally conceived to provide voice services, most older switching and transmission equipment is optimized for operation with this type of signal. When circuits that have been designed to provide analog voice services are utilized for data transmission, devices called modems must be utilized. A *modem* (MOdulate–DEModulate) is a device that takes the digital signal from data communications device and translates it into an analog signal that is compatible with the public network. Because of the narrow bandwidth of the analog network, analog transmission of digital signals is generally limited to a digital speed of 9600 bits per second (bps) or less.

## Digital Services

A *digital* signal is defined as one that is discontinuous in nature and changes from one form to another in discrete steps. The telecommunications needs of the digital computer have placed pressure on telephone companies to provide telecommunications services of much higher speed and greater reliability than are available with analog circuits. Although telcos have been using so-called "digital" circuits (actually digitally encoded analog circuits) for many years to transmit calls between central offices, digital services to end users is a relatively recent offering. Digital circuits offer many advantages over traditional analog signaling techniques. Because the facilities accept digital signals directly, no modem is required between the data communications device and the public network (although in most cases some form of electronics is still required to interface to the system). Moreover, several digital signals can be multiplexed or combined onto a circuit supported on one or two pairs of wires, greatly increasing the information-carrying capacity of the physical cable plant.

Digital services are currently available in most urban areas to support both voice and data services. Digital data services are offered at a number of speeds, from 2400 bits per second (2.4 kbps) up to 1.5 million bits per second (Mbps). Although digital speeds greater than 1.5 Mbps are available and in use by the telcos for interoffice circuits, 1.5 Mbps is generally the maximum speed that can be provided from a central office to a user location over existing copper cable plants.

Through various coding techniques, a voice conversation can be encoded onto a channel running at a speed of 64 kbps. While sophisticated compression techniques can code speech at speeds as low as 16 kbps. speech quality at this slow speed becomes very poor, and the 64-kbps coding technique has become an internationally recognized standard for voice digitization. This 64 kbps rate forms the basis of a *digital multiplexing hierarchy*, that is used for the transmission of both data and digitally encoded voice signals throughout the public network. The next step in this digital hierarchy is a combination of twenty-four 64-kbps signals, through the use of time-division multiplexing, into a single 1.544-Mbps bit stream known as a *T-1* or *DS-1 signal*. This T-1 rate is fast becoming the most popular of the digital services, because it is available to end users on the same copper pairs as are traditional voice services, and because of the wide availability of equipment that supports this transmission speed. A T-1 signal can be utilized to transmit any combination of 24 voice and data circuits from a user's premises to a central office, where the data stream can be transmitted intact to a new location or can be broken up and the various information "packets" routed to different destinations.

Above the T-1 level, multiple T-1 facilities can be further multiplexed to levels where a single high-speed circuit [operating at 139.264-Mbps (DS-4) rates] can support up to 4032 individual voice circuits simultaneously. Because these speeds can only be supported on expensive coaxial or fiber-optic cables, and due to the tremendous amount of electronic equipment required to perform the multiplexing, these high-speed services are generally not available to individual subscribers but are reserved for use by the telephone company as interfacility trunks.

The T-1 facility and the digital multiplexing hierarchy have become so universally accepted for digital transmission that they form the basis for most currently planned future telco service offerings. These formats are also utilized as the basic building blocks for terrestrial microwave and satellite transmission systems.

## PLANNING CONSIDERATIONS

The three most important factors in planning an urban telecommunications infrastructure are (in order of priority) flexibility, flexibility, and (most important) flexibility. The one thing that can be learned from the problems that have been encountered in trying to support more and faster circuits on our existing cable infrastructure is that forward-looking planning and space for future expansion are critical if newly installed cable plants are to support the telecommunications requirements of a metropolitan area for any length of time.

Because the design of facilities such as central offices are generally the domain of the local telephone company, there is little, beyond providing good locations and adequate space, that an urban planner can contribute to the construction and utilization of these facilities. However, in the area of cable plant distribution, good planning can contribute a great deal to the overall effectiveness of the telecommunications infrastructure.

In most urban areas, modern telecommunications cable plants are distributed from the central office to subscriber locations through an underground network of conduits and duct banks. Because these duct banks have a finite volume available for the distribution of cable, sufficient space must be provided within the system to support both initial requirements and the inevitable future expansion. It must also be taken into account that the local telco will not be the only user of this duct bank system. It is quite possible that in addition to telecommunications services, this system will be utilized to support such other services as cable television, private networks (both fiber and copper based), and the services of a telecommunications carrier other than the local telephone company.

A critical part of the telecommunication distribution system is the cable vault. It is here that cables are spliced and large multipair cables are broken down into multiple smaller cables for distribution into individual buildings. Traditionally, cable vaults have supported only copper cables, and as such did not have rigid environmental requirements. However, the proliferation of fiber-optic cables within the telecommunications infrastructure has had an effect on the construction of the vault. Because the quality of splices in a fiber-optic cable has a great effect on the quality of transmission, these splices must be made under closely controlled conditions of temperature, humidity, and cleanliness. Also, in some cases the cable vault must support the electronic equipment necessary to convert the multiplexed fiber-optic signal back to its original copper form for distribution to subscribers. This equipment also has very stringent requirements for power, cooling, and environmental controls. Due to these strict requirements, a controlled environmental vault (CEV) is generally recommended for use in telecommunications distribution systems that will utilize large amounts of fiber-optic cable.

The CEV is usually equipped to support both the fiber optic cable and its associated electronics, and the necessary termination and cross-connect hardware for the copper conductors. In planning for the use of CEVs it should be remembered that CEV is considerably larger than a traditional cable vault and has sizable power requirements to support both the electronic equipment that it houses and the necessary environmental control systems.

## EMERGING TECHNOLOGIES

Over the next few years, almost all local and long-distance telephone companies will be undertaking capital improvement programs that are designed to support their customers well into the twenty-first century. These programs are intended to put in place systems and equipment that can provide a wide variety of voice, data, and video services to a variety of subscribers ranging from the largest corporations to the smallest residential users. It is planned that these services will utilize much of the existing copper and fiber-optic cable plant, and as such may not have much direct impact on the planning and installation of an urban telecommunications infrastructure. However, because these services have the potential to effect profoundly the way we do business, they will certainly have other, far-reaching effects on the urban environment.

### Integrated Services Digital Network

The integrated services digital network (ISDN) is the latest attempt to standardize the transmission of voice and data services. As discussed previously, voice and data transmission systems have, until now, utilized a wide variety of speeds and transmission formats to communicate with other systems. This lack of standardized transmission systems has hampered the exchange of information over the public network. Systems could exchange information only if they transmitted at the same speed, utilizing the same transmission format or protocol. Further complicating matters, analog circuits had limited data-carrying capacity, and some digital data circuits could not handle voice. ISDN is intended to change this situation.

In its simplest form, ISDN establishes a digitally encoded circuit, at 64 kbps, as the basic telecommunications service available to subscribers from the central office. As you may recall, this 64 kbps is the minimum digital circuit speed necessary for voice communications. All other, higher-speed circuits (for use in data transmission) will be integer multiples of this rate.

The so-called ISDN *basic rate interface* is known as 2B + D. This means that a single circuit supported on one pair of wires will contain two B (for "bearer") channels each operating at 64 kbps, and one D (for "data") channel running at 16 kbps. Each B channel can support either single-voice conversations or data communications traffic at a rate up to 64 kbps. The D channel will support control and signaling information that is required by the system. (Until recently, control and signaling information for a digital circuit was supported within the data channel itself. This meant that if a circuit was operating at 64 kbps, something less than that amount of useful data could be transmitted. The remainder of the channel was given over to signaling information. Under ISDN, however, the signal and control information is carried on the D channel, allowing the full bandwidth of the B channel to be utilized for data or voice services.) For small users, one or more basic rate circuits could support all of their voice and data communications needs. Basic rate service, on individual copper pairs, is envisioned for most residential and light commercial users.

The *primary rate interface* (23B + D) is envisioned for most large-scale commercial users. This service consists of 23 bearer channels and a single data channel, all operating at 64 kbps multiplexed onto a single T-1/DS-1 circuit, either copper or fiber. Primary rate service will support multiple voice and data circuits to a customer's premises and will require some form of on-premises electronics to convert the single 1.544-Mbps signal into multiple 64-kbps signal. By the time ISDN is ready for widespread use, most PBXs will have the capability to interface directly to the primary rate service, disassemble the signal into its constituent information packets, and distribute these packets to their appropriate destinations.

The most promising aspect of ISDN is its standardization of digital communications services. In its full implementation, which may be some years away, ISDN offers worldwide end-to-end digital communications for a wide variety of voice, data, and video services, all on the same physical plant. Combinations of these services can be transmitted to the local central office as a single high-speed signal. At the CO, this signal can be disassembled into its constituent information packets and each packet forwarded to a different destination.

## Very Small Aperture Terminal

In our previous discussion of satellite systems, we made mention of the fact that the considerable size (30 feet or more in diameter) and expense of satellite earth stations limited their use to organizations that could both afford them and had the space available to locate them. This second point is most important in an urban area, where space is at a premium. Developed to work in conjunction with several of the newer, higher-powered satellites, very small aperture terminals (VSATs) function in a manner similar to the larger satellite dishes, but are several times smaller. Typically, VSATs can be as small as 5 to 6 feet in diameter, which makes them suitable for use in urban areas where the relatively large open spaces that are required for traditional satellite antennas are not available. Additionally, their light weight and small profile make them ideal for mounting on the roofs of buildings without having to reinforce a building's structure as may be required for the installation of the larger conventional systems. When VSATs become commercially available, users who formerly could not make use of the advantages that satellite telecommunications systems offer will finally be able to enter this market.

## Video Teleconferencing

Because of its high bandwith requirements, video transmission has traditionally been excluded from the list of regular services provided by the local telephone company. Until recently, a subscriber who wished to have services such as video teleconferencing had to bear the cost of installing an expensive coaxial-cable bypass network, or purchase a microwave or satellite system for video services. However, the advent of high-speed services such as ISDN, coupled with the increasing use of fiber-optic cables, has made the transport and delivery of video services by the local telco possible.

Most people are familiar with audio teleconferencing, which allows groups of people in widely separate locations to carry on a discussion or meeting through one or more telecommunications connections as if they were all in the same room. This is simply a more sophisticated version of a conference call. The addition of video to this process can add greatly to its overall effectiveness. The use of full-motion video not only allows the participants to view each other but allows slides, videotapes, film, and other graphic materials to be used in the course of the meeting, greatly increasing its effectiveness and productivity. Although a full-scale video teleconferencing center is extremely expensive (on the order of $100,000 or more), smaller portable setups will soon be in the price range that most organizations can afford.

## TELECOMMUNICATION STANDARDS

In recent years, the development of large-scale, cost-effective telecommunications networks has been hampered by the lack of existing standards. Traditionally, each computer system manufacturer has developed proprietary hardware and software that best suits the needs of its equipment. This has led to our current situation, where in most cases, systems from different manufactures cannot "talk" to each other. This problem has been compounded by telecommunications systems manufacturers taking the same approach to the development of transmission and switching systems. The end result of this process is a situation where, even if computers by the same manufacturer need to exchange information, they will be able to do so only if they utilize telecommunications equipment from the same or compatible manufacturers.

It has recently become obvious to most computer and telecommunications systems manufacturers that this situation cannot be allowed to continue. For technologies such as ISDN and video conferencing to become feasible, all users must utilize the same or compatible transmission techniques. In the United States, the Institute for Electrical and Electronic Engineers (IEEE) has undertaken the development of a series of telecommunications standards, of which ISDN is one, that are intended to replace or augment all the existing incompatible standards. In Europe a similar process, based on the same standards that have been proposed by the IEEE, is currently being conducted by the CCITT, which is a major European standards body.

Most of the standards being addressed by both the IEEE and the CCITT are related to the logical techniques of telecommunications transmission and affect the systems software and hardware. A parallel effort, sponsored in the United States by the Electronics Industries Association (EIA), is designed to establish standards for the copper and fiber-optic cabling that will interconnect these systems. While these standards processes are proceeding quite slowly, with much lobbying by certain equipment manufacturers to establish their transmission techniques and equipment as the "standard," it is hoped that eventually, techniques and protocols acceptable to all will be established. If these organizations are successful in doing this, the telecommunications revolution that will occur as a result will make our current revolution pale by comparison.

## FUTURE IMPACTS

In the very near future, many practical applications of the services and equipment described in this chapter will begin to have noticeable impacts on our daily lives. Many of us are already intimately familiar with such products as cellular telephones and fax machines, and most of us could not dream of living without our answering machines. The effects of these changes are already being felt in the planning community, and architects, engineers, and planners have begun to consider the telecommunications factors in the early stages of the design process.

Some of the more obvious accommodations that have to be made to the telecommunications explosion have to do with physical planning of commercial buildings to support these systems (Fig. 5-1). In addition to larger risers, more flexible methods of distribution, and appropriate power and cooling within individual buildings, the overall telecommunications infrastructure of our cities and suburban areas has to be updated to support these technologically enhanced structures.

Certain technologies however, may have more subtle effects. Videoteleconferencing, for instance, may ultimately reduce both long- and short-range travel. Cost-effective video teleconferencing could help to reduce business travel by allowing participants in different cities to conduct meetings from their own offices or from a local teleconferencing center, rather than wasting time traveling hours or days to conduct meetings face to face. If small, self-contained units that use standard telecommunications circuits (such as the ISDN described previously) become available, intracity traffic patterns could be affected as business people conduct "crosstown" videoteleconferences rather than traveling to an associate's or client's office.

Postal services could also be affected by the growth of such services as electronic mail, facsimile systems, and advanced data communications networks. Computer Aided Drafting and Design (CADD) systems have become a great tool for architect and engineers, reducing substantially the time required to produce or revise construction documents. But "hard" copies must still be produced and hand carried to the client's office by mail or messenger. The introduction of high-speed public data networks will permit these drawings to be transmitted almost instantaneously anywhere in the world, eliminating the need to produce and deliver paper documents. Fax machines are already being used to send letters, memos, and sketches to their destination, and electronic mail systems are being utilized by most major corporations to reduce both intraoffice and interoffice mail volumes. This trend shows no sign of slowing down or reversing. E-mail, electronic banking, and interactive video systems will all contribute to an overall reduction in traditional mail services in residential communities as well. Soon, any home with a computer terminal, a telephone line, and a television set will have all they need to shop for clothes and food, pay their bills, and even browse through the latest catalogs.

A good example of a public institution that will be affected by the advances in this field is the local public library. Academic institutions are already begin-

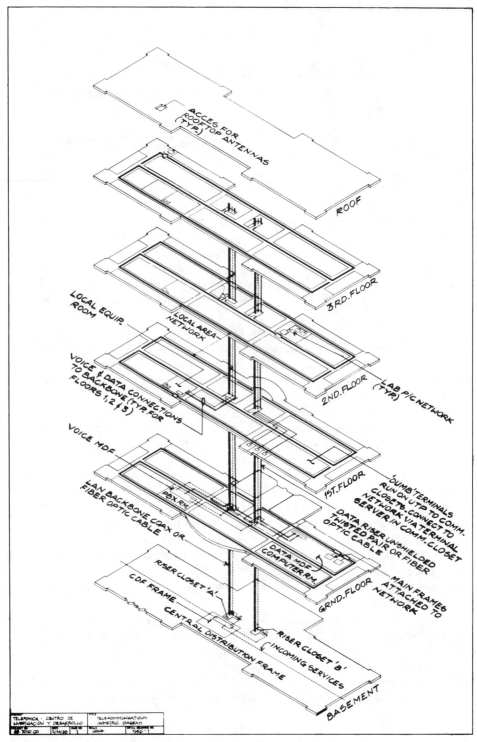

**Figure 5-1** Building plan for support of telecommunication system. (From Flack & Kurth.)

ning to utilize high-density mass storage devices and data communications networks to provide on-line electronic data bases for student and faculty use. These services allow users to access vast amounts of written and graphic information, regardless of their location or the time of day. When these services are made available to the public in general, library operating hours could be greatly curtailed as people start utilizing the facilities electronically.

It is difficult to predict the long-range impact of these technologies on the way that we conduct our lives. We can feel certain that cost-effective, user-friendly telecommunications equipment will be available to a wide range of commercial and residential users in the near future. However, our society has many deeply rooted traditions, especially within the business world. The equipment and services described in this chapter have the capacity to change fundamentally the ways in which we have done business for hundreds of years. It may ultimately be society's resistance to these forces of change, more than technologic limitations, that determines how long it will take before this revolution occurs.

**Additional Reading.** Because the planning and construction of a telecommunications infrastructure has until very recently been the providence of the telephone company, very little information is readily available to the architect or planner. The information that is available to the design community is generally written for practitioners in the field and is full of technical jargon and information not useful in addressing overall planning issues. Happily, this is beginning to change. Following is a partial list of organizations (and some of their publications) that make available to the general public information useful in the process of planning telecommunications systems:

1. *Building Industry Consulting Services International (BICSI).* BICSI is a trade organization that addresses issues of importance within the telecommunications industry. They generally focus their attention on the physical aspects of telecommunications, and have been instrumental in promoting the development of standards and practices in the field. Their *Telecommunications Distribution Methods Manual* offers guidelines for planning telecommunications infrastructures, with particular emphasis on the practical aspects of cable plant design.

2. *Datapro Research Corporation.* Datapro Research publishes a series of manuals that address a wide variety of issues in the field of telecommunications. These documents, which are periodically updated for subscribers, cover most of the current and emerging technologies and are a good source of quick, concise information on a wide range of topics.

3. *Business Communications Review.* In addition to a monthly periodical on telecommunications topics and issues, BCR conducts 2- to 3-day short courses covering most telecommunications topics. While these courses have traditionally covered topics, such as telecommunications protocols and software, that are not of any interest to the urban planner, BCR has recently introduced courses on local area networks, cabling system design, and bypass technologies, all of which should be of interest to the designer. The cost of the course generally includes a course manual, which can be a valuable reference source long after the course is over.

4. *The Urban Land Institute.* The Urban Land Institute has published a two-volume set entitled *Buildings and Technology: Enhanced Real Estate* that covers issues related to the design and marketing of technologically enhanced or "smart" buildings. These books provide valuable information about this topic and should be read by anyone contemplating the design of this type of structure.

## QUESTIONS

1. What was the single most important factor contributing to the delay in modernizing the U.S. telecommunications infrastructure?

2. Describe the telecommunications environment in the United States just prior to the breakup of AT&T.

3. How has divestiture affected both users and providers of telecommunications services?

4. What are the advantages of a PBX to a large commercial telecommunications user?

5. What factors have limited the growth and modernization of the telecommunications infrastructure in existing urban areas?

6. What technological solutions have been utilized to overcome these limitations, and why have they developed slowly?

7. Discuss some of the currently available telecommunications services, and explain the factors that contributed to their development.

8. Explain what is "bypass technology? Give some examples and discuss their benefits to the telecommunications user.

9. What are some of the factors that must be considered in planning a modern telecommunications infrastructure?

10. Describe some of the emerging telecommunications technologies, speculate on the impact that they will have on society, and discuss their meaning in terms of the planning process.

Route 9, Ossining, New York

# 6

# STREETS

**Ekkehart R. J. Schwarz**
Schwarz & Zambanini P.C. Architecture,
Engineering, Urban Design

# INTRODUCTION

Streets appear to be a simple part of the infrastructure, yet their planning and design is a complex process requiring the expertise of several engineering, planning, and design professions: the highway, traffic, and electrical engineer; the environmental, transportation, and city planner; and the urban designer, architect, or landscape architect. The highway engineer designs new streets and prepares reconstruction plans for existing streets to ensure that they stay in good repair and are able to carry the daily load of traffic. The traffic engineer is responsible for the flow of traffic, to ensure that streets are laid out in such a way that the traffic flows as smoothly as possible and grid locks do not happen too often. The electrical engineer has to assure that streets are properly lit, providing safety for the motorist and the pedestrian, and that lighting does not consume too much energy.

The environmental, transportation, and city planners' responsibility is to develop plans showing the relationship of streets to the rest of our environment. They try to ensure that the traffic does not poison us or cause other harm, that a balance between different modes of transportation is found, and that the amount of traffic on streets is kept at a reasonable level. They can develop plans relating the number and width of streets in a rational way to the number of buildings serving them. One of their most difficult tasks is to develop plans assuring that the traffic created by the large amount of space in new buildings constantly being added to the bulk of our cities can be carried by the existing street network.

Finally, it is the task of the urban designer, the architect, or the landscape architect to give form to streets and public spaces. The engineering and planning aspects will determine many parts of a project. But it is the architectural design professional who can turn an engineering project into a special design— who is trained to create a design making a public place, a local street, a park, or neighborhood plan unique, and in the best cases, a work of art.

This brief chapter can only give a short introduction into the many fields of expertise necessary to accomplish good street design. The reader looking for more detailed information is advised to consult the "Additional Reading" listed at the end of this chapter. Most authors include extensive bibliographies for their field of study.

# HISTORICAL OVERVIEW

Streets are the oldest form of infrastructure. As soon as human beings left their early villages and settled in cities, we find streets lined by houses, along which all transportation and communication is organized. The form the road takes from one house to the next becomes a manifestation of the way people communicate with one another.

Streets are connected to form large networks. They are the basic elements in the plans of settlements, villages, and cities. Most street networks have grown over centuries without a plan, but some have been laid out in a plan at one point in time. The street plan of a city gives a very vivid picture of its history and its different areas of use. In New York, for example, the downtown natural growth of the early Dutch and British settlement was followed by the rectangular grid of the commissioners' plan of 1811.

*Grown street networks* are the result of settlements whose development has taken a long time and for which no large-scale plans have been made. Early Egyptian, Mesopotamian, or Greek settlements, medieval cities, and early American towns are all examples of grown street networks. Streets form loosely connected irregular grids. They open at times to larger open areas, often places of particular importance: a church, city hall, or palace. Streets may be curved to follow the topography or have a sudden change in direction.

*Planned street networks* follow a plan laid out at a specific time, in most cases by a governing authority. We find them early as well: housing for Egyptian workers, Greek colonial cities, and city foundations in the Roman empire; Renaissance cities, new city foundations in the North and South American colonies, cities of the European baroque. Most planned street networks for cities are based on a *rectangular grid*. Sometimes the axes are emphasized as major streets and some blocks are left open for plazas in front of major buildings or parks.

A special form of the grid, the *radial grid*, was developed during the baroque, to focus on the overwhelming importance of the seat of the ruler, the palace as focal point. Karlsruhe in Germany is a famous example of such a plan. The radial grid was used for the plan of some American cities, such as Washington, D.C., and Savannah, Georgia.

The emergence of the automobile after the industrial revolution made necessary the separation of uses of streets and of land. Zoning was invented to separate use of the land into industrial, commercial, and residential areas, and new types of streets were created. Highways became "streets" for cars only, and streets used by cars and pedestrians began to have sidewalks.

New forms of street grids were used to accommodate new classes of streets and to help the separation of areas of different use. Highway networks began to overlap street networks, and a new form of a street network was created, the treelike network. The *tree network* is a hierarchical form of network which allows access from only one direction, like the branches of a tree. In street layout the shortest distance between two points is no longer as important as the environmental quality of the residential neighborhood. People driving home can easily drive a little farther to a quiet home without taking too much additional time.

Many plans of new towns and suburban neighborhoods have some form of mixture of a grid and a hierarchical tree network.

The typical rectangular street plan of many American cities is an example of a planned grid network, whereas the hierarchical plan used in many modern suburbs is an example of a treelike network. The intersections and exchange points between these different types of grids, from a signalized street intersection to large highway intersections, have become the symbol of twentieth-century transportation networks.

## PURPOSE AND USE OF STREETS

Streets are part of the public realm of our communities. Most streets are public, for use by everybody, serving private land. In addition to their primary use for pedestrian and vehicular transportation, they provide space for many other parts of the infrastructure, such as water and sewer lines, electricity, gas, telephone, and cable television in their right-of-way above or below ground.

People use streets for many other activities: as places to walk and talk, for recreation and play. Many commercial activities take place on and along streets. Goods are displayed in many different ways: streets are lined with store windows, merchandise is displayed in front of stores, kiosks are located at strategic points, peddlers offer their goods moving from location to location. Occasionally, streets are used for special temporary activities such as parades, fairs, markets, or demonstrations.

Research and analysis of streets that are well used and preferred by people has shown that the livelihood of a street depends greatly on a well-balanced mixture of vehicular traffic and pedestrian activities. Donald Appleyard has studied the impact of traffic direction and volume on the different uses of residential streets and their influence on family life. The layout and traffic volume of streets can be a decisive factor in people's decision to stay or move from a neighborhood.

In addition to the balance of uses on residential streets, a city needs a variety of different street types to make districts lively and interesting. Kevin Lynch lists a variety of prototypes of streets: the boulevard, the freeway, the parkway, the pedestrian promenade and the pedestrian shopping street, curving suburban streets, and cul-de-sacs. Special street types can have very significant influence on neighborhoods: Freeways through cities have in most cases a negative influence, whereas well-designed boulevards are a great asset. New Yorkers are again discovering the great beauty of Ocean and Eastern Parkways in Brooklyn, and much present neighborhood revitalization is centered around such special types of streets.

# THE STREET SYSTEM

## Space Allocation for Streets

Streets are designed to provide a route for transportation between two points and to carry an estimated volume of vehicles and pedestrians. The relationship between the land used for streets and the land they serve depends on the type and density of development. Low-density detached housing on separate lots requires longer streets than does higher-density attached cluster or multifamily housing. Higher density, though, causes a higher volume of traffic. More traffic requires more streets. But in a high-density area one can imagine a point where more land than is available will be required for transportation. In such cases, other means of transportation have to be found or traffic comes to a standstill. It is the task of land use and transportation planners to find the right balance between space needed for vehicular and pedestrian movement and space allocated for other uses.

Forecasts for additional traffic can be obtained from a study of existing and proposed traffic originators. Both vehicular and pedestrian traffic volumes are determined by them. They are mostly buildings, but they can be sport fields, airports, or beaches as well. Every new originator adds to the volume of traffic and often has a negative impact on the traffic flow of an area. Environmental impact statements are used to determine such effects and to propose measures of mitigation.

Many streets designed today are a reconstruction of existing streets with a given right-of-way. They are reconstructed because they are in disrepair or the use of the street has significantly changed. The volume of vehicular and pedestrian traffic changes for several reasons. Zoning parcels may be built to their legally allowed density, or higher densities are granted as a variance from existing zoning or as a bonus for providing special amenities. People change their travel behavior, the number of trips, or the means of transportation. As a result, the volume of traffic exceeds the capacity that streets can comfortably provide.

The space available in a given right-of-way for pedestrian and vehicular traffic has to be reallocated. A compromise between street space used for moving lanes, turning lanes, and parking lanes, and space used for bike lanes and sidewalk, has to be found. How best to use the available street space often leads to heated debates between residents, community boards, and municipal agencies, and requires a delicate political decision-making process.

## Street Classification

Streets within communities are classified as arterial, collectors, and local streets. Some streets are designed for special uses such as malls and boulevards.

**TABLE 6-1   CHARACTERISTICS OF STREET CLASSES**

| Item | Street Class | | | |
| --- | --- | --- | --- | --- |
| | Freeway | Arterial | Collector | Local |
| Average trip length | Over 3 mi | Over 1 mi | Under 1 mi | Under 1/2 mi |
| Average travel speed | 50 mph | 25–45 mph | 20–30 mph | 25 mph |
| Access control | Full | Partial | Partial | Minimum |
| Spacing | 2 mi | 1 mi | 1/2 mi | 300–500 ft |
| Linkage | CBD and major generators | Secondary generators | Local areas | Land parcels |
| Traffic volume, (ADT) | 50,000–100,000 | 15,000–50,000 | 2000–15,000 | 100–2000 |
| Traffic control | Free flow | Stop signs on cross streets | Stop signs on cross streets | Must stop or yield |
| Percentage of VMT | 0–40% | 40–70% | 10–20% | 5–10% |

Source: JHK Associates, Design of Urban Streets, U.S. Dept. of Transportation. Federal Highway Administration.

Many "streets" are restricted to a specific type of traffic. Highways, parkways, expressways, and throughways are for use by vehicles. Walkers and bikers find "ways," walkways or bikeways, in many locations, designed especially for them. The different classes of streets are distinguished by their characteristics in trip length, travel speed, access control, spacing, traffic volume, traffic control, the points they are linking, and the percentage of overall traffic they carry. Table 6-1 shows data typically used by traffic engineers.

### Downtown Transportation Plans

Downtown areas of large cities need customized transportation plans to accommodate the large amount of vehicular and pedestrian traffic. San Francisco has developed a transportation plan that shows transit preferential streets, primary vehicular streets, and proposed commuter bike streets. A supplemental plan for a pedestrian network shows existing and proposed exclusive pedestrian walkways, part-time pedestrian streets, pedestrian/service streets, and pedestrian-oriented vehicular streets and arcades.

# DESIGN OF STREETS

The street designer, a civil engineer specializing in street design, has the prime responsibility for a street construction project. His or her work is based on research and plans of transportation planners and the designer incorporates the work of traffic and electrical engineers and of architects or landscape architects into the contract. At the beginning of a project, surveys of the physical environment and the existing traffic are taken and an analysis of the existing or planned traffic is prepared. Then all other agencies or companies that might want to add work to the contract—the water and sewer departments, the utility telephone and cable-TV companies—are contacted. The bearing quality of the soil and the condition of existing utilities and of all other parts of the infrastructure the street will carry have to be examined. Community groups and local planning boards are consulted to determine the impact of the design.

## The Surveys

The survey of the physical environment has two parts, the topographic and utility surveys. The *topographic survey* shows all existing elements above ground. It is taken by a licensed surveyor and is a complex document locating all buildings and building lines, curbs, walks, manholes, catch basins, lights, signals, traffic signs, parking meters, police and fire call boxes, mailboxes, fences, and planting and trees. The *utility survey* shows all utility lines and structures above and under ground. In high-density areas it is a complex document showing all sewer, water, electricity, gas, steam, telephone, cable-TV, and any other utility lines. It may also show the location of underground structures such as extended basements or vaults, subway tunnels, and pedestrian underpasses and tunnels.

The *traffic survey* is performed by a traffic engineer to collect data for design decisions. Traffic counts are taken either manually or by machine to measure existing traffic characteristics, such as overall traffic volumes, percentages of trucks and buses, number of turning movements, and the level of pedestrian activity. All three surveys are used to determine the alignment and the grades of the street and to locate new utilities, lighting, and trees.

## Traffic Analysis and Capacity

Traffic analysis uses data collected during the traffic survey and a projection of future traffic to determine the capacity a street should have. Since the infrastructure has a useful life of 20 to 30 years, existing traffic characteristics are usually projected into the future to ensure that growth in traffic can be accommodated. The capacity of streets is measured separately for vehicles and pedestrians. Vehicular traffic flow is measured in vehicles per lane per hour during the peak hour and compared to its theoretical capacity. This analysis leads to the *v/c ratio,* which is an indication of how well a facility is operating.

*Level of service* (LOS) is another useful method of describing operating characteristics of facilities. It is classified from LOS A to F. LOS A of service represents a free flow at high speed, LOS B and C stable flow and decreasing speeds, LOS D lower speed and congestion, LOS E unstable flow, and LOS F forced flow. At signaled intersections the level of service designations are based on the average stopped delay per vehicle. For example, a LOS E at a signalized intersection corresponds to a delay of 40.1 to 60.0 seconds per vehicle.

Capacity and the level of service can be improved by the installation of traffic lights, the use of traffic control agents, the provision of additional travel lanes for turns, wider traffic lanes, restricting heavy vehicles, lessening the grade of the road, restricting turns, and computerizing traffic signals. Capacity and safety of a street system can be improved by the use of a one-way street system, or lock-free intersections, to prevent *gridlock.*

Pedestrian traffic flow is measured in persons per square foot and flow rate. It is described in level of service as well. For example, a LOS C describes a condition where 24 to 39 square feet per pedestrian with expected speed greater than or equal to 240 feet per minute and a flow rate less than or equal to 10 pedestrians per minute per foot width are available (*Highway Capacity Manual*).

In many cases, Fruin's LOS concept for pedestrians has been used, which uses 15 to 25 square feet at an average speed of 10 to 15 PFM (pedestrians per foot of effective walkway width per minute) as acceptable space per person at a level of service C for walkways. Queuing standards (people waiting on platforms or in front of elevators) are 7 to 10 square foot per person with an average interperson spacing of 3 to 3-$\frac{1}{2}$ feet at a level of service C. The level of service here ranges from level A through F. If the space is available, it can be improved by widening sidewalks or by the use of moving walks and escalators, measures that are more readily taken in buildings. In streets, sidewalk obstructions such as signposts, street posts, or newspaper dispensers can be removed from queuing areas as well as areas opposite display windows or entrances to buildings.

## Right-of-Way

The width of a right-of-way is a function of several factors: the proposed vehicular and pedestrian capacity, anticipated land use, topography, cost, and most important, the function of the particular street in the street network hierarchy. JHK Associates recommends 80 to 130 feet for arterial, 60 to 80 feet for collectors, and 50 to 70 feet for local streets. An important consideration in the width of the right-of-way is the desired travel speed of vehicular traffic, which is influenced by the lane width since vehicles in narrow lanes tend to proceed more slowly.

## Lane and Roadway Width

Expressway standards utilize 12-foot-wide lanes to accommodate high speeds and wide vehicles and trucks. Since parkways often prohibit wide vehicles, 11-foot lanes are generally acceptable. On urban streets 10- to 11-foot lanes are acceptable. Eight feet is most often used for parking lanes and 4 to 6 feet for a bike lane. The use of parking lanes for bus stops increases the widths to 10 feet, and use of the parking lane for bus-only lanes during certain hours specifies 10 to 11 feet. Heavy bus and truck traffic calls for an additional foot; special snow removal or drainage conditions may warrant an additional 1 to 2 feet. Lanes should not be wider than 16 feet, to avoid misuse by drivers as a second lane. The roadway width varies from 36 feet for a local road with two parking lanes to 86 or 92 feet for a six-lane road with a 22-foot median and no parking.

Medians can be painted or raised physical barriers separating opposite directions of traffic. A median with left-turn bays or two-way left-turn bays does considerably increase the safety of operation. Physical medians vary from 4 feet for narrow barriers to 14 feet for medians with turn bays and 24 feet for wide medians. Painted medians with left-turn bays should be 10 feet, and with two-way left turns, 12 feet.

## Intersection Design

Intersection design is a key element in street design. Capacity and travel time are significantly influenced by its design. The basic types of intersections are three-leg ("T" intersections), four-leg and multileg intersections, rotaries, and interchanges. Rotary intersections are commonly used in Europe on large plazas around monuments, as in Trafalgar Square in London and the Siegessäule in Berlin. They are suggested for use under moderate traffic conditions and often are upgraded using channelization and traffic signal controls as traffic volumes increase. Interchanges employ ramps and grade separations to improve traffic flow. They are expensive and used most frequently on freeways.

Design considerations for an intersection are turning radii for vehicles from 25 feet for passenger cars to 46 feet for a semitrailer/full trailer combination. Left- and right-turn lanes require special consideration. Left-turn bays should be installed on arterials if 20 to 50 left turns per hour occur and there are insufficient gaps in the opposing traffic stream. A left-turn bay should have a minimum median of 14 feet and a minimum taper of 60 feet.

Traffic signal controls are installed when their installation is justified by warrants. At the present time the MUTCD lists 11 warrants:

1. Minimum vehicular volume
2. Interruption of continuous traffic
3. Minimum pedestrian volume
4. School crossings
5. Progressive movement
6. Accident experience
7. Systems
8. Four-hour volume
9. Peak-hour delay
10. Peak-hour volume
11. A combination of warrants

## Roadway Design and Drainage

The horizontal and vertical alignments are determined by planning requirements and terrain conditions, by design speed and sight distance, and by horizontal and vertical curve design. A car moving at design speed can be accommodated without problems by a road designed for that speed. The design speed should be about 10 miles per hour greater than the running speed or speed limit. The sight distance is the stretch of roadway visible to the driver. The minimum sight distance should be the distance the driver needs to react and bring the car to a safe stop.

*Horizontal curve design* determines the most desirable curve a street should have. The design factors for a horizontal curve are superelevation, side friction factor between the tires and the road, and vehicle speed. *Vertical curve design* determines the minimum and maximum grades that a street should have. The minimum grade is normally 0.5%, to allow for proper drainage, and the maximum 5 to 11%, depending on topography and design speed.

Drainage of streets is designed for the largest rainstorm in 10 to 50 years: 10 years for curb inlets on arterial streets and 50 years for depressed roadway sections and main underground drain systems. The main concern for the street designer is the flow of water on the surface of the street. The water should be moved as quickly as possible and not become a safety hazard for the pedestrian, bicyclist, or motorist. The typical cross-slope of an urban street is 2 to 4%, sloped toward the gutter (Fig. 6-1). On inverted crown streets, slopes toward a median are possible.

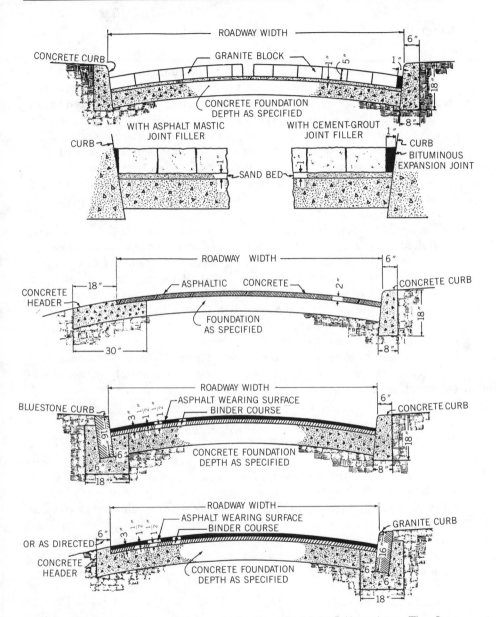

**Figure 6-1** Typical cross section for streets. (From DeChiara & Koppelman, *Time Saver Standards for Site Planning*, McGraw Hill, New York, pp. 654-655, 1984.)

## Street Lighting

Light sources for roadway lighting are incandescent, fluorescent, and high-intensity discharge (HID). Incandescent and fluorescent light sources are rarely used in new installations because of their inefficiency. Four types of HID lamps are available: *Mercury vapor* has a bluish-white color and is used on many American roadways. *Metal halide* is more efficient and has better color rendition. It is often used around sport stadiums. *High-pressure sodium* has a gold-white color and is most often used in new installations in the United States. It is very efficient and has a long lifetime. *Low-pressure sodium* has a deep yellow color. It is the most efficient lamp but has poor color rendition. It is used more commonly in Europe.

The level of illumination is measured in footcandles or lux. Lux is the metric equivalent of footcandles. The recommended level depends on many factors: type of street, use of the area, and roadway surface. Residential collector roads, for example, have a recommended level of 0.6 footcandle with a uniformity ratio of 4:1 to 3:1. Every light source has a distribution curve with maximum footcandles in the middle and a reduction of footcandles to a point of half-footcandles on various curves.

The mounting height is typically 30 feet, sometimes up to 50 feet. At intersections or large open areas it can reach 100 to 150 feet. The spacing is 70 to 200 feet either staggered on both sides of or placed separately on each side of the street.

## Traffic Signals, Signs, and Markings

The JHK manual lists a number of advantages and disadvantages of traffic signals. Their advantages are orderly movement of traffic, reduction of certain types of traffic accidents, interruption of traffic flow to allow other traffic to enter or cross, increase in capacity of side streets, greater cost-effectiveness than manual controls, and driver confidence. Their disadvantages are increased intersection delay, increase in rear-end accidents, reduction in capacity of the main street, and unnecessary delays caused by unwarranted or poorly operated signals.

Traffic signals are operated by controllers. There are many different types in use. They can be retimed or traffic actuated by devices close to the intersection, traffic adjusted through remote control, and/or operated by minicomputers or microprocessors. In most cases traffic is detected by inductive loop detectors, magnetometer detectors, or magnetic detectors. Other detectors—pressure-sensitive detectors, radar detectors, and sonic detectors—are also in use. Pedestrian and bicycle signals can be operated manually by pushbuttons.

Other important decisions in signal design for an intersection relate to the *phasing*, the number of independent movements through an intersection; the *sequence*, the order in which the phases occur; and the *cycle*, the number of seconds it takes to complete one cycle. Intersections can be designed for two up to eight phases. The number of phases increases the safety but decreases the capacity of the intersection.

Traffic signs and markings are formalized in the *manual on Uniform Traffic Control Devices for Streets and Highways*. The manual is the official guideline for signs and markings on federal-aided highway systems. Control devices serve three purposes: regulation of traffic, warning of roadway conditions, and guiding of traffic along certain routes.

# LANDSCAPING OF STREETS

The landscaping of streets, including choice of materials, street furniture, and planting, has a lasting impact on the appearance and use of a street. Trees, paving materials, street furniture, and special lighting are the most commonly used elements in a street landscaping plan. In many jurisdictions the community has a strong impact on the extent of landscaping of the streets. Two questions are frequently discussed with community groups. Is the community willing to take some responsibility for the maintenance of landscape items? Many jurisdictions are willing to pay for the capital cost of special items only if the community is willing to participate in their maintenance. Second, will undesired people take over and make the redesigned space their home? Many people prefer no street improvements if they feel that these improvements will make the street a hangout for drug dealers and the homeless.

## Trees and Other Plantings

One of the cheapest and most effective ways to improve the appearance of a street is to plant trees. They can be treated as architectural elements and used to emphasize a street corridor or special place or be used as a focal point. They should be planted in the ground and not in planters, and only varieties that have been proven to survive on streets, "street trees," should be used. Different varieties of trees can be used to identify special streets and places, such as "cherry tree lane" or "maple street." Shrubs and flowers could be planted in malls or in planters at special locations.

## Paving Materials

Another effective way to improve the quality of a street is through the choice of materials for sidewalks and curbs. Most American sidewalks are made of concrete. Different-color concretes can be used to provide variety. Bricks, cobblestones along the curb, and granite curbs are beautiful sidewalk materials, especially when they are used to complement the surrounding architecture. Bluestone was widely used in the past and is still used to provide a sense of history in landmark districts. Other natural stones, especially granite and travertine, are materials used on sidewalks around public buildings or high-rise structures to provide a lasting material and a distinct environment.

A 5-foot strip along the curb paved with bricks or cobble stones can be quite effective. All trees, signs, fire hydrants, parking meters, newspaper vending machines, and mailboxes can be located within this strip so that the strip serves as a natural barrier between the roadway and the street.

## Street Furniture

Benches and seats, tables, and wastepaper baskets make a landscaped area more attractive and invite people to use public areas. In his studies William H. Whyte has shown how people use outdoor spaces. They like to sit where they can watch other people and what's going on in the street. They avoid places that are isolated, difficult to reach, or hostile. The cost of maintenance is an important factor in selecting street furniture. The ideal bench does not need maintenance and can easily be replaced. Using standard designs is economical and allows for easy replacement of damaged pieces.

## Special Lighting

Lighting can be used to create a distinctive appearance. Instead of normal "cobra head" lights placed in equal intervals of about 50 to 70 feet on alternate sides of the roadway, special lighting fixtures can be placed opposite each other or following another special design plan. Street lights can emphasize the direction of a street or areas of importance. Used in such a way they become an element of architectural design. Fixtures of special design, for example nineteenth-century cast-iron fixtures, are frequently used for such areas.

Streetlights are often positioned to illuminate the roadway and not the sidewalk. In urban areas, the level of sidewalk illumination should be considered. Pedestrian areas not reached by regular street lights, and walkways under trees and along waterfronts and in parks and plazas, need special pedestrian lighting. Such fixtures are only 10 to 15 feet high and create an atmosphere of comfort and safety for pedestrians.

## Leftover Street Space Used for Pedestrian Amenities

The foremost purpose of the public right-of-way is to provide space for transportation. Nevertheless, within any given network of public right-of-ways space exists, often left over from the space needed for the actual roadway or sidewalk, that can be used for public amenities (Figs. 6-2 to 6-6). These small sitting areas or parks provide the opportunity for an improvement in public life, especially in low- or middle-income neighborhoods where people need to escape their small, crowded apartments. The decision as to how many of these leftover spaces to provide is made during the alignment or realignment process of a street.

It takes the skill of an experienced urban designer with knowledge of traffic planning to negotiate with traffic engineers an alignment that permits the creation of usable leftover space. Some short service roads are not needed; traffic islands can be enlarged or sidewalks extended to create space for additional amenities without an adverse impact on smooth traffic flow.

# PLANNING ISSUES

## Life Cycle of Streets

Most street and highway construction done today is actually reconstruction of an existing street network. The life cycle of New York City streets, that is, the time after which the average street is reconstructed, is approximately 30 years. Many streets, though, have not seen any work for much longer times. Some streets have lasted longer, but many have deteriorated due to community neglect in maintaining streets.

Today, municipalities as well as states and the federal government face large street and highway reconstruction programs. This is a great chance to employ some of the design tools that have been developed and to address planning issues with which we are constantly confronted, such as downtown traffic and transportation plans, urban design guidelines to coordinate private and public activities, plans for pedestrian and bicycle networks, improved plans for coordination between street reconstruction and utility construction, and improved plans for citizen participation.

## Downtown Plans

Some cities, including Denver and San Francisco, have developed comprehensive downtown plans. Other cities, including New York City, have developed their downtown on a piecemeal basis using special transportation plans, urban

before

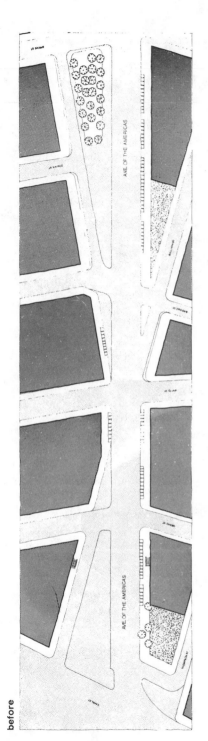

Avenue of the Americas Reconstruction

after

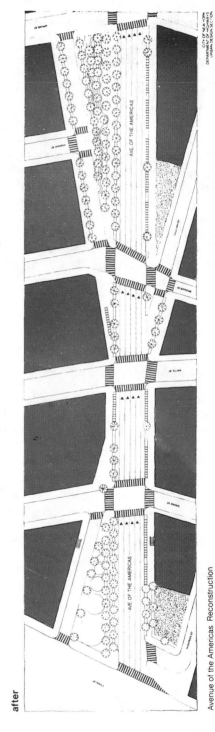

**Figure 6-2** Avenue of the Americas between Canal and Spring Streets before and after reconstruction in 1976. During the alignment discussions, substantial areas of leftover street space could be gained for pedestrian use.

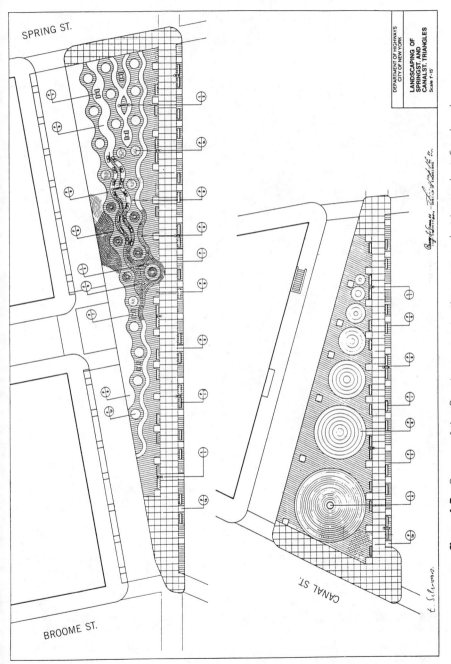

**Figure 6-3** Avenue of the Americas reconstruction: new pedestrian park at Canal and Spring Streets. Two-color bricks were used in imaginative patterns to create paving that matches the surrounding architecture.

SPRING ST.

BROOME ST.

CANAL ST.

DEPARTMENT OF HIGHWAYS
CITY OF NEW YORK

LANDSCAPING OF
SPRING ST. AND
CANAL ST. TRIANGLES
Scale 1"-15'

**ABINGDON SQUARE**
AXONOMETRIC

0 ▬▬▬▬ 100 F

**OFFICE OF URBAN DESIGN**
NEW YORK CITY DEPARTMENT OF TRANSPORTATION
BUREAU OF PLANNING AND RESEARCH

**Figure 6-4** Proposal for realignment of curbs at Abingdon Square in Manhattan. A new pedestrian area will be created (the "nose" with the column) separating head-on one-way traffic.

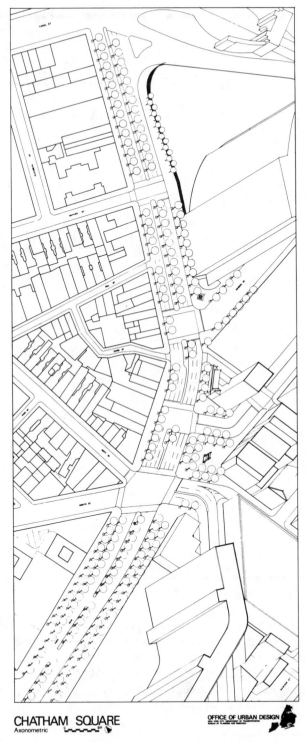

CHATHAM SQUARE
Axonometric

OFFICE OF URBAN DESIGN

**Figure 6-5** Proposal for redesign of Chatham Square in Manhattan. The new alignment will create a smoother traffic flow and a pedestrian plaza in the heart of Chinatown.

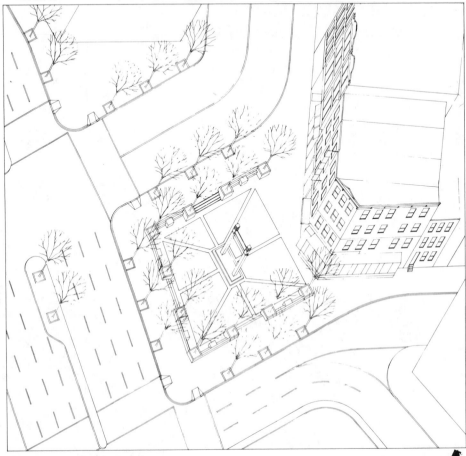

# CHATHAM SQUARE

OFFICE OF URBAN DESIGN
NEW YORK CITY DEPARTMENT OF TRANSPORTATION
BUREAU OF PLANNING AND RESEARCH

**Figure 6-6** The proposed plaza for Chatham Square in Chinatown uses an existing war memorial as the central piece and surrounds it with trees and benches. At the present time the memorial stands on a small island surrounded by traffic.

design guidelines for special areas, and special district zoning. Without a plan it is difficult to coordinate the many actions of private developers and municipal departments. Although a comprehensive plan for New York City has proven incomprehensible and therefore of little consequence, a plan for a smaller area such as downtown or midtown could specify different street types and their role in the development of a defined area.

## Integration of Highways into Urban Areas

Highways are designed for the exclusive use of vehicles traveling at high speeds. Although the construction of large highway networks has created enormous freedom of movement between cities and to their suburbs and beyond, it has often had a negative impact on the immediate environment. Construction of large highway networks can lead to the destruction of large parts of existing neighborhoods and can dislocate their residents. The need for faster transportation has to be weighed against the need to leave communities and the natural environment intact. Many incomplete or aborted highway projects, such as the Embarcadero in San Francisco and the West Way in New York City, are testimony to the fact that satisfactory integration into the city could not be found. If the integration of a highway into an urban neighborhood is part of an overall development plan and well-designed, satisfactory solutions, such as F.D.R. Drive under the United Nations building and the Brooklyn–Queens Expressway under the Brooklyn Heights promenade, can be found.

## Pedestrian and Bicycle Networks

In many American cities pedestrians and bicyclists have to fight for their right-of-way on streets. Yet we know that it is the pedestrian who makes a street lively and desirable. We have accumulated a wealth of knowledge about pedestrian behavior, space requirements, and preferences. We have built many malls and special places for pedestrians. But we are just beginning to develop plans that outline pedestrian and bicycle networks and integrate them into development plans.

## Urban Design Guidelines

Urban design guidelines are an excellent tool to use to give an urban street its form. The appearance of a street is determined by the design of the buildings along its building line. Buildings can form a large variety of spaces, street rooms, street walls, and plazas. The appearance of buildings is influenced by many zoning regulations, air and light standards, bulk and height regulations, special district regulations, and design guidelines. Guidelines are designed to help developers and architects make buildings that relate to one another and make the street a coherent whole. They guide design and development without knowing or determining the exact detail or result of the design. Streets can benefit immensely from well-written guidelines.

As an example, see the Design Guidelines for the South Residential Area of Battery Park City prepared by Cooper Eckstut Associates in 1981 (Figs. 6-7 to 6-9). They are based on a master plan developed in 1979 which determines site

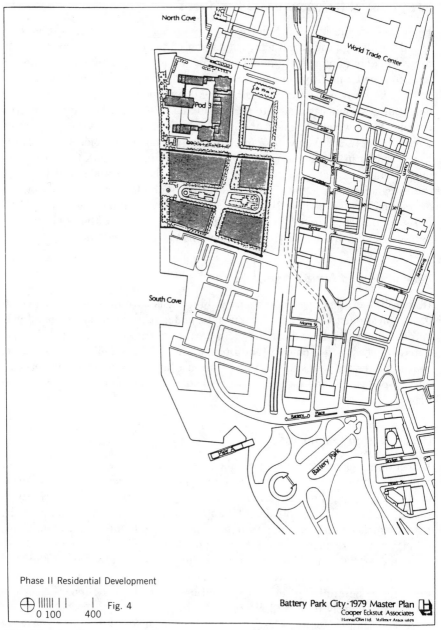

Phase II Residential Development

⊕ ||||| | | | Fig. 4
0 100    400

Battery Park City · 1979 Master Plan
Cooper Eckstut Associates
Hanna/Olin Ltd.  Vollmer Associates

**Figure 6-7** South residential area of Battery Park City in lower Manhattan. (From Cooper Eckstut Associates.)

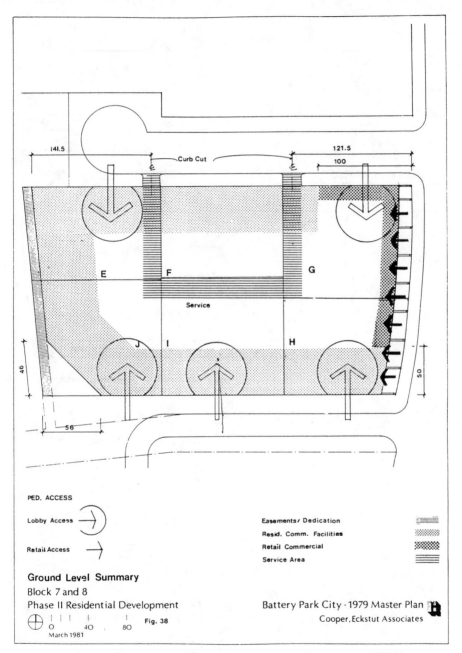

141.5
121.5
Curb Cut
100

E    F    G

Service

J    I    H

46

50

56

PED. ACCESS

Lobby Access →

Retail Access →

Easements/ Dedication
Resid. Comm. Facilities
Retail Commercial
Service Area

**Ground Level Summary**
Block 7 and 8
Phase II Residential Development
0    40    80    Fig. 38
March 1981

Battery Park City · 1979 Master Plan
Cooper, Eckstut Associates

**Figure 6-8**  South residential area of Battery Park City: example of the 1979 Masterplan requirement. (From Cooper Eckstut Associates.)

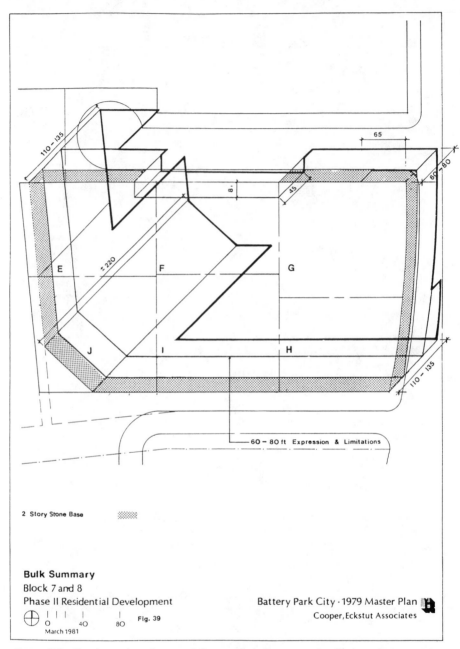

110 - 135

65

60 - 80

8.

45

≈ 220

E

F

G

J

I

H

110 - 135

60 - 80 ft  Expression  &  Limitations

2  Story Stone Base

**Bulk Summary**
Block 7 and 8
Phase II Residential Development

Battery Park City · 1979 Master Plan

Cooper, Eckstut Associates

0        40        80        Fig. 39
March 1981

**Figure 6-9**   South residential area of Battery Park City: example of urban design guidelines.  (From Cooper Eckstut Associates.)

conditions, circulation, and bulk. The guidelines require architectural elements to be used throughout (signage, lighting, exhaust, balconies, projections, materials and colors, incremental heating and cooling units, landscaped roofs, arcades, the esplanade wall, and fences) and determine separately for each parcel use and access and details of the bulk (density, coverage, street walls, and building heights).

## CONCLUSION

This chapter has served as a short review of the issues that must be addressed during the planning and design of streets. A close working relationship between the professionals involved—planners, engineers, and architects on the one hand, and the people and their governments on the other hand—is needed to create successful streets.

Streets represent the public realm of our built environment. Just as buildings are a symbol of the private structure of our society, streets are a symbol of its public structure, a symbol of the way we relate to one another. Skyscrapers represent the power of our corporations, and single-family homes represent the dream of freedom and independence of our citizens. What do streets say about our society?

This century has seen large advances in the design and construction of buildings and highways. But the connection, the point where buildings and highways meet, the street, has seen little advance. Only lately have we begun to think about streets again: the role they have in public life, the way buildings relate to them, and the way they are integrated into the built environment of our cities. The next great task in design of our streets will be to match the advances we have seen in the design of buildings and highways. We have to design streets that are livable again and part of an environment that is safe and which we can all enjoy.

## ADDITIONAL READING

### History of Streets

Benevolo, Leonardo, *The History of the City*, MIT Press, Cambridge, Mass., 1980.
Moholy-Nagy, Sybil, *Matrix of Man: An Illustrated History of Urban Environment*, Praeger, New York, 1968.

## Purpose and Use of Streets

Appleyard, Donald, *Livable Streets*, University of California Press, Berkeley, 1981.

Lynch, Kevin, *A Theory of Good City Form*, MIT Press, Cambridge, Mass., 1981.

## The Street System

Department of City Planning, City and County of San Francisco, *The Downtown Plan*, The Department, San Francisco, 1983.

JHK Associates, *Design of Urban Streets*, U.S. Department of Transportation, Federal Highway Administration, Technology Sharing Report 80-204, Washington, D.C., 1980.

## Design of Streets

Federal Highway Administration, *Manual on Uniform Traffic Control Devices for Streets and Highways*, FHWA, Washington, D.C., 1978.

Fruin, John J., *Pedestrian Planning and Design*, Metropolitan Association of Urban Designers and Environmental Planners, New York, 1971.

JHK Associates, *Design of Urban Streets*, U.S. Department of Transportation, Federal Highway Administration, Technology Sharing Report 80-204, Washington, D.C., 1980.

Pushkarev, Boris S., with Jeffrey M. Zupan, *Urban Space for Pedestrians: A Report of the Regional Plan Association*, MIT Press, Cambridge, Mass., 1975.

Transportation Research Board of the National Research Council, *Highway Capacity Manual*, Special Report 209, TRB, Washington, D.C., 1985.

## Landscaping of Streets

Lynch, Kevin, *Site Planning*, MIT Press, Cambridge, Mass., 1984.

Project for Public Spaces, Inc., *Managing Downtown Public Spaces*, New York, 1984.

Whyte, William H., *The Social Life of Small Urban Spaces*, Conservation Foundation, Washington, D.C., 1980.

## Planning Issues

Alexander, Christopher, Hajo Neis, Artemis Anninou, and Ingrid King, *A New Theory of Urban Design*, Oxford University Press, New York, 1987.

Anderson, Stanford, ed., *On Streets*, Institute for Architecture and Urban Studies, MIT Press, Cambridge, Mass., 1978.

Barnett, Jonathan, *Urban Design as Public Policy: Practical Methods for Improving Cities*, Architectural Record Books, New York, 1974.

Cooper Eckstut Associates, Battery Park City South Residential Area Design Guidelines, prepared for the Battery Park City Authority, 1981, in *Urban Design: Theory and Practice*, conference papers by the Hartford Architecture Conservancy, 1984.

## QUESTIONS

1. What is the difference between grown and planned street networks?

2. Name primary and other activities occurring on streets.

3. How are streets classified and what are the characteristics of each class?

4. Name the different surveys taken at the beginning of a street design project.

5. What is the "level of service" in traffic analysis and what are the different classes used?

6. What is horizontal and vertical alignment in street design?

7. How can a community influence the design of a street?

8. Name four major elements of street landscaping.

9. What are left over spaces in street design and how can they be used?

10. What are urban design guidelines? Give an example.

The George Washington Bridge

# 7

# BRIDGES

**Samuel I. Schwartz**
**Joseph DePlasco**
NYC Department of Transportation

# INTRODUCTION

*On the night of April 12, 1988 a group of city, state, and private consulting engineers assembled in the drab muster room of the 7th Police Precinct on Pitt Street in New York City. They had just spent the afternoon examining floor beams, stringers, and columns under the Manhattan approach of the Williamsburg Bridge.*

*The mood of the meeting was somber. As the engineers discussed the situation, a subtle tension suggested that the pending decision was inevitable but nonetheless unthinkable. Once the country's largest suspension bridge, the Williamsburg was to be closed because the safety of the traveling public could not be ensured.*

*The closing traumatized neighborhoods on both sides of the river, shook the city administration, and became a national emblem of the neglect of our infrastructure. Presidential candidates referred to the bridge in speeches, and one even trekked across the still open pedestrian way during the New York primaries.*

Each year more than 100 bridges collapse in the United States, killing approximately 12 people. Most of these bridges are relatively minor and receive little national attention. The Williamsburg Bridge (Fig. 7-1) did not collapse, no one was killed or injured, yet the closing attracted national and even international attention. Part of the reason was that the Williamsburg Bridge is in New York City, the media capital of the world, and it is a major road and rail crossing that connects an interstate highway with the country's largest business district. But there was more to the story. The public had heard about infrastructure neglect—but never before were so many people affected in an instant. Furthermore, a prevailing belief is that what happens in New York City is a precursor of events elsewhere. As we point out in this chapter, a similar story can be told throughout the country.

There had been many warnings. In September 1967 the Silver Bridge over the Ohio River collapsed, killing 46 people; the nation was shaken from its complacent attitude. The federal government responded by creating a national inspection program and increasing capital spending for bridges, as explained below. However, more than 20 years later it is clear that not enough has been done.

The National Council on Public Works Improvement issued a report to the President and Congress in 1988 entitled, *Fragile Foundations: A Report on America's Public Works.* As part of the report, the Council prepared a report card for the country's public works, including bridges, and gave them a grade of "C−"—barely a passing mark. What is perhaps most telling, according to the Federal Highway Administration, is that 41.2% of the country's bridges were

**Figure 7-1**  The 85-year-old Williamsburg Bridge in New York City was closed for three and a half months in 1988 after engineers detected deteriorated steel and severe corrosion.

rated either structurally deficient or functionally obsolete as of June 1988 (Federal Highway Administration, 1989).

How did we get to this point? There were two great eras of bridge construction in the United States. From the late nineteenth century through the first half of the twentieth, waterway bridges were constructed at an unprecedented level as connections for urban centers and as rail links. Many of them continue to serve the public. Following World War II, the country undertook the largest highway construction program the world had ever seen. Forty percent (by area) of the country's bridge decks were constructed during the 1950s and 1960s.

The early waterway bridges were built with great strength and many redundant parts. The cities that built them also provided for their maintenance and assigned a large number of tradespeople to care for them. The post–World War II bridges are sleeker, have a lower safety margin, and have fewer redundant elements. As they were built, greater attention was placed on openings, while maintenance needs languished. The echo generated from the Williamsburg Bridge closing was the thundering silence of the inevitability of this kind of neglect.

## THE GREAT ERAS OF BRIDGE BUILDING

*To me, a bridge is more than a thing of steel and stone; it is the embodiment of the effort of human heads and hearts and hands.*

—*D. B. Steinman, 1939*

### Waterway Bridges

At the turn of the century, bridge engineers were deemed romantic heroes by their contemporaries. They were the equivalents of today's astrophysicists, yet unlike modern-day scientists, bridge builders created tangible and visible structures that had an immediate and dramatic impact on people's lives.

In 1912, Willa Cather used a bridge engineer, Bartley Alexander, as the protagonist in her novel *Alexander's Bridge.* He was a handsome, rugged man with technical expertise, who nonetheless failed to comprehend the limitations of his science. The novel, which was based on the collapse of the Quebec Bridge in 1907, ends with the tragic death of the bridge builder and the collapse of his greatest project.

Science reached new heights during these years. Cities crept upward into the sky and bridges were almost yearly built longer and higher than the ones before them. This development marked a change in how and by whom the urban environment was constructed. The country's earliest bridges, built in the seventeenth century, were wooden structures used to span canals and creeks. In New York City in the 1650s, Dutch settlers built wooden bridges over canals in lower Manhattan. The first waterway bridge in New York City connecting Manhattan to the mainland was constructed over Spuyten Duyvil Creek in 1693. Kingsbridge was a simple structure consisting of wood and stonework. Nonetheless, it served New York City for nearly 220 years.

Designs for more complicated bridges both in the eighteenth and nineteenth centuries were frequently developed not by engineers, but rather, by artisans who were in the forefront of the burgeoning scientific community. Thomas Paine, the English pamphleteer who participated in the American Revolution, worked on a design for an iron bridge for nearly two decades. Similarly, Thomas Pope, a New York City craftsman and carpenter, designed a bridge for the East River in 1810. If built, Pope's Flying Pendant Lever Bridge would have spanned 1800 feet and climbed to a height of 223 feet above the water.

By the early nineteenth century a convergence of science and technology set the groundwork for breakthroughs in bridge construction. American and British engineers were becoming familiar with earlier bridge construction practices in China and Latin America. Written works began to appear in Europe and North

America on different bridge construction techniques, including James Finley's 1810 article, "A Description of the Patent Chain Bridge" and Thomas Pope's *A Treatise on Bridge Architecture* (1811) (Kemp, 1984).

Bridge designers were experimenting with a large variety of bridge types. The purpose remained the same: to span from one piece of land to another. But it was soon realized that a bridge need not be limited to spanning water. As cities expanded in the nineteenth century, land bridges (or viaducts)—a crossing over a road or railroad tracks—became necessary. In the 1860s the landscape architect Frederick Law Olmsted and his colleague Calvert Vaux designed and built the first land bridge over another road in the United States, in Central Park in New York City. The land bridge revolutionized vehicular and pedestrian traffic by vastly increasing intersection capacity and improving safety.

Railroads transformed both the environment and bridge building. Between 1830 and 1880 rails crossed the nation, requiring bridges for rivers, canals, and canyons. John Roebling completed the country's first unbroken rail link between New York and Chicago with his Niagara Falls Suspension Bridge (1855) (Martin, 1984).

The largest nineteenth-century railroad bridge was built in Scotland. The Firth of Forth Rail Bridge opened in 1890, at $1\frac{5}{8}$ miles long, replacing the Brooklyn Bridge as the world's longest structure. It took seven years to construct the Forth Bridge. Unlike many earlier rail bridges, the Forth Bridge superstructure was constructed entirely of steel—nearly 54,000 tons. Steel was still a relatively new substance in the late nineteenth century. In fact, steel was not made in the United States until 1858. The first bridge constructed of steel in America was the 1874 Eads Bridge over the Mississippi River at St. Louis.

It was the widespread use of steel that modernized bridge building in the nineteenth century. With the expansion of the railroad system, iron rail bridges had to be rebuilt at an astonishing rate because they either collapsed or became obsolete due to heavier traffic. The rate of failure—nearly 40 a year—along with the loss of life, led George Vose to publish a book in 1886 entitled *Bridge Disasters in America*. In the mid-1850s the Bessemer process of making steel was developed, and steel quickly became the material of choice for bridges.

The Roebling family pioneered the use of steel in bridge construction. John and Washington Roebling's Brooklyn Bridge (1883) was the first suspension bridge to use steel cabling. Before looking at the various kinds of bridges, it is worth spending some time on John Roebling because the lives of Roebling and his son Washington are so connected to the history of bridges in the United States.

John Augustus Roebling was born in Germany in 1806. In 1823, Roebling left home and enrolled in the Royal Polytechnic Institute of Berlin. There he found himself caught up in liberal intellectual circles and soon enrolled in classes taught by the great German philosopher Hegel. Roebling became dissatisfied with the Prussian state and developed a fondness for America.

Like other young engineers at the time, Roebling's interests went beyond math and science. His reading spanned philosophy, the arts, and the natural sciences, leading him to an almost spiritual belief in our ability to overcome nature. This idea was captured in his presentation on the proposed Brooklyn Bridge to the New York Bridge Company in 1867:

> *The contemplated work, when constructed in accordance with my design, will not only be the greatest bridge in existence, but it will be the greatest engineering work of this continent, and of the age. Its most conspicuous features, the towers, will serve as landmarks to the adjoining cities, and they will be entitled to be ranked as national monuments [quoted in Reier (1977)].*

Before he presented his proposal for the East River Bridge, as the Brooklyn Bridge was then known, Roebling had worked in several fields and had already completed a bridge over the Niagara Falls and one in Cincinnati. While working along the Harrisburg–Pittsburg rail line as a surveyor, he noticed that the hemp rope used to pull barges frequently broke. Remembering an article he read as a student, he set up a shop to make wire rope by twisting the iron wire into cables (Reier, 1977).

Roebling first used his cables on the suspended aqueducts he built over the Alleghany River, proving that his wire rope was indeed stronger than the iron chain used in suspension bridges built earlier.

Roebling perfected his wire ropes on a number of projects before he turned to the Brooklyn Bridge. However, before he could begin construction on the bridge he died as a result of injuries caused during an accident while he was surveying for the location of the Brooklyn tower. He passed the task of constructing the bridge on to his son Washington. The younger Roebling turned to the bridge with the same enthusiasm and dedication as had his father.

The Roebling family's involvement with the Brooklyn Bridge did not stop with Washington. While overseeing construction of the bridge, Washington fell ill. The chambers below the river used to build the support foundations for the towers, called caissons, had to be constructed under water. Because of the depth of the caissons, workers could stay below for only a few hours, and even then, they were susceptible to "caisson disease," known today as "the bends." In 1872, Washington Roebling had to be carried out of the Manhattan caisson. Although he lived through the construction of the bridge, he remained an invalid and had to watch the progress of the work from his home in Brooklyn. Emily Roebling, Washington's wife, oversaw the remaining stages of construction.

## Land Bridges

The great era of land bridges had its origins in the late nineteenth century. As mentioned earlier, Frederick Law Olmsted and Calvert Vaux built the country's first land bridges in Central Park in the 1860s, allowing for different modes of transportation to pass under and above one another without interference. In the 1880s and 1890s bicycle enthusiasts led the movement for "speedways" (hard, smooth-surfaced roads) for recreational use.

The country's first limited-access and landscaped motor parkway, the Bronx River Parkway, underwent construction in 1907. The Bronx River Parkway was a divided highway that made extensive use of overpasses and entrances and exits so that motor cars could travel unimpeded by local traffic. The idea behind the limited-access highway is to avoid intersections and traffic signals. Each time a highway has to cross another road or railroad, a bridge must be constructed over the highway or the highway must bridge the other road.

The growth of America's public roads was slow in the early years of the twentieth century. It was the federal government's involvement that spurred the construction. Under the New Deal Works Progress Administration (WPA) in the 1930s, some 651,000 miles of roads were built (Patton, 1986).

Road building was interrupted by World War II but was resumed with even greater vigor in the 1950s. In 1954, President Eisenhower appointed a special committee to solve the country's road problems. The Interstate system that was developed was marketed as a national defense program that would connect all parts of the country. Soon a driver could travel from Maine to Florida or from Boston to San Francisco without stopping at a traffic light. Today there are nearly 44,000 miles in the Interstate system, 1.1% of the 3.9 million miles of roads and streets that tie the country together. There are now over 270,000 bridges in the federal-aid system (Patton, 1986; Highway Users Federation, 1988).

## BRIDGE TYPES

A simple bridge is one that is supported at two points, whether it is a log across a stream or a steel structure resting on two piers. In addition, there are basically six common types of bridges: suspension, cable-stayed, arch, movable, truss, and concrete. The cable-stayed and concrete bridges are largely phenomena of the twentieth century.

### Suspension Bridges

Of all the bridge types that we depend on daily, the suspension bridge has captured the greatest interest. Although the suspension bridge appears complicated, it is one of the oldest bridge forms, probably dating from around 100 B.C. The first suspension bridge in America, constructed with iron chains, opened in Uniontown, Pennsylvania, in 1796.

Although building methods have advanced, the basic principle behind the suspension bridge remains the same. Generally, two to four cables are draped over two towers some distance apart. The distance between the towers is called the *main span*. The cable ends are embedded in huge concrete and masonry blocks to resist the pull of the weight of the bridge (known as the *dead load*) and the weight of the traffic (the *live load*). Smaller wire cables, known as *suspenders*, hang down from the main cables to support the roadway. Figure 7-2 shows a basic suspension bridge, with main cables, suspenders, eyebars, anchors, and towers.

When John Roebling designed the Brooklyn Bridge, aerodynamic theory (wind motion) was not as advanced as today. Although Roebling used trusses (a type of steel framing described later), which add rigidity and limit movement caused by wind and weight, he depended more on the use of stays, diagonal cables that run from the towers to the bridge roadway. Roebling first began to think about stays while on a ship to America (Hindle, 1984). In studying the sails, he observed that the stays held the masts in place. He used stays on all but one of his suspension bridges.

### Cable-Stayed Bridges

The cable-stayed bridge is among the most popular long-span bridge designs since World War II. The principle is simple. A large tower is built and diagonal cables emanate either from the top of the tower (fan type) or from intervals along the tower (harp type) to support the roadway. The stays allow for the use of stiffer decks and, uniquely, combine the suspension design with a concrete deck. The advantages over a standard suspension bridge include speed of construction and cost, since anchorages are not necessary. There are also no massive cables, as with suspension bridges, making cable repair or replacement simpler.

### Arch Bridges

Like suspension bridges, arch bridges have historical precedents (Fig. 7-3). False arches were constructed at least 3000 years before the start of the Christian Era. It is called a false arch because unlike true arches that are held together by compression, the stones are balanced on top of one another. True arches date back to between the fourth and eighth centuries B.C. The Romans perfected arches and used them for aqueducts and highways. Stone arch bridges remained the design concept choice until the eighteenth century.

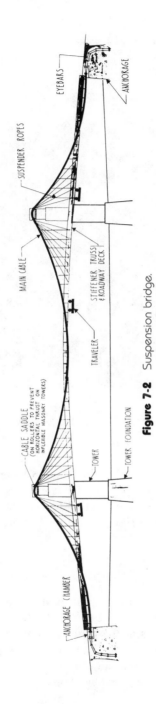

**Figure 7-2** Suspension bridge.

**Figure 7-3** The Bayonne Bridge (Arch Bridge).

The Washington Bridge (1889) over the Harlem River in New York City is a good example of an arch bridge. The bridge consists of two 510-foot steel and iron arches. Both spans have six arched girders composed of individual steel plates riveted together. The Washington Bridge was the first arch bridge in the United States to use plated girders.

## Movable Bridges

Across many waterways it was either impractical to too costly to build bridges with high clearances. Nonetheless, the waterways had to be crossed. There are four types of movable bridges in the country. *Swing bridges* are supported on central piers and are rotated horizontally, on either a pivot or a turntable. The swing bridge has the advantage of not limiting the height of a passing vessel, but it does restrict the horizontal clearance. *Bascule bridges*, otherwise known as *drawbridges*, are probably the oldest movable bridges. Modern bascule bridges are normally of the two-leaf type, where the leaves open by means of an axle and counterweight. However, there are some single-leaf varieties. *Vertical lift bridges* are useful when the horizontal clearance required is greater than the vertical. The entire span of the bridge is lifted by cables directly along the towers of the bridge. The *retractile* or *sliding bridge* was popular at the beginning of the twentieth century but has since fallen out of favor. The Carroll Street Bridge in New York City is the oldest of the four remaining retractile bridges in the nation. To allow for boat passage, the deck rolls back on rails onto the adjacent land.

## Truss Bridges

The first long-span truss was constructed across the Ohio River in 1864. In the late nineteenth century, steel trusses became popular for railroad bridges, although they remained essentially similar to earlier wooden trusses. A truss bridge consists of several members, either wood or metal, joined together in a series of triangles. The live and dead loads are distributed along the beams in such a way that no beam takes a disproportionate share of the load. The type of truss depends on the configuration of the triangles. Truss members are placed either in tension, whereby the members pull on one another, or in compression, where they push on one another from both ends. Figure 7-4 shows a large truss bridge. The truss has remained popular in the twentieth century, although more recently it has been substituted for reinforced concrete slabs, T-beams, and steel plate girders.

## Concrete Bridges

In the late nineteenth century, engineers began to experiment with concrete reinforced with steel. Many of these bridges were similar in design to arch bridges. More recently, reinforced concrete has been used with girders, which are solid beams that extend across the span. With the development of the highway system in the late twentieth century, steel and concrete girder bridges became one of the most popular bridge designs. Figure 7-5 shows a typical highway bridge. Although there are a wide variety of bridges, many of the components are either the same or similar for each. There are two areas that should be looked at briefly: the substructure and the superstructure.

The *substructure* consists of those elements (abutments and piers) that distribute the load of the bridge to the ground below. *Abutments* are walls of reinforced concrete or masonry that support a bridge's superstructure and approach roadway and retain the embankment. *Piers* are the intermediate supports for a multispan bridge. They are normally composed of steel or reinforced

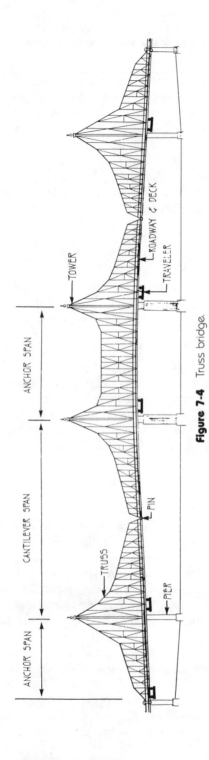

**Figure 7-4** Truss bridge.

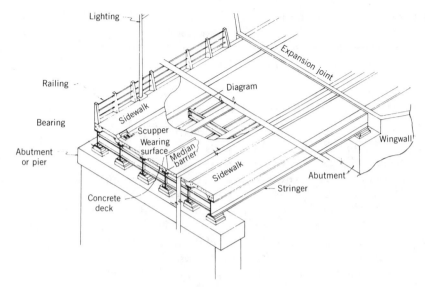

**Figure 7-5** Highway bridge.

concrete and can be either walls or columns. The *superstructure* consists of all those elements that are supported by the abutments and piers. The bridge deck supports the traffic load on the bridge and passes it on to the floor system. The floor system is normally made up of transverse floor beams and longitudinal beams called *stringers* that are supported by widely spaced floor beams.

## NATIONAL BRIDGE PROGRAMS

On December 15, 1967 the Silver Bridge over the Ohio River collapsed. Of the 37 vehicles on the bridge, seven fell to the ground below and 24 plunged into the river. Forty-six people died and nine were injured. The Silver Bridge opened in 1928. Although it looked like a standard suspension bridge, the main cables were constructed of individual *eyebars*, which are long bars with an eye or hole at the end. The eyebars were then bolted together to make something that resembled a chain. One hundred and forty-eight eyebars were used. After the bridge collapsed, salvage crews discovered one eyebar that had fractured. The National Transportation Safety Board concluded that the eyebar had failed because of time-dependent factors: stresses, corrosion, and fatigue. It was the worst bridge disaster of the century (Mair, 1982).

Immediately following the disaster, a presidential task force was formed to determine the reasons for the collapse and to study the existing bridge inspection program. Two bridge safety programs were introduced as a result of the task force's work: the National Bridge Inspection Program and the Special Bridge Replacement Program.

## National Bridge Inspection Program

The National Bridge Inspection Program (NBIP) was established by the Federal Aid Highway Act of 1968. Standards were developed for the inspection of federally aided bridges. Federal-aid systems are those routes that are deemed of greatest importance to the nation. As laid out in the Federal Highway Administration's *Status of the Nation's Highways and Bridges: Conditions and Performance* (1989), they (1) connect major metropolitan areas, cities, and industrial centers; (2) serve the national defense; and (3) connect border points with routes to Canada and Mexico. There are 3.87 million miles of public streets and highways in the nation. Nearly 850,000 miles of the national highway network is designated as federal-aid. NBIP set forth three standards for these routes:

1. Methods for highway departments to use in conducting inspections.
2. Minimum time lapse between inspections
3. Qualifications for the people who carry out inspections

With July 1, 1973 set as the target date, all states were required to inventory and inspect their federal-aid bridges on a biennial basis. The Surface Transportation Assistance Act of 1978 extended the program to off-system bridges. The initial inventory and inspections were to be completed by December 31, 1980. Today, every state has a biennial inspection program.

## Special Bridge Replacement Program

The Special Bridge Replacement Program was created by the Federal-Aid Highway Act of 1970. The program was designed to supplement the states' efforts to replace unsafe bridges. The act authorized $100 million for fiscal year 1972 and $150 million for 1973. In 1978 the Surface Transportation Assistance Act expanded SBRP to what is now known as the Highway Bridge Replacement and Rehabilitation Program (HBRRP).

To determine the condition of a bridge, SBRP created a sufficiency rating system that uses a numerical rating between 0 and 100 on 84 items relating to safety, serviceability, and essentiality for public use. Structures with a rating of less than 50 become eligible for federal funding for replacement. The sufficiency rating is determined by the following formula:

$$\text{sufficiency rating} = S(1) + S(2) + S(3) + S(4)$$

where

$S(1)$ = structural adequacy and safety

$S(2)$ = serviceability and functional obsolescence

$S(3)$ = essentiality for public use

$S(4)$ = reductions for special deficiencies

HBRRP allows bridges with a rating of 80 or less to be eligible for rehabilitation rather than replacement. The new program also greatly increased federal funding, with $4.2 billion authorized for the four fiscal years 1979–1982. However, the program includes bridges off the federal-aid system, as well as bridges over highways and railroads. Since its inception HBRRP has been used to fund improvements to 28,714 bridges. Between fiscal years 1979 and 1987 over $11 billion has been spent (Federal Highway Administration, 1989).

## NATIONAL BRIDGE CONDITIONS

*While the [Schoharie Creek] bridge's design and construction may well have been deficient, with proper inspection and maintenance, this bridge would not have collapsed.*
—**Report of the New York State Commission of Investigation, 1987**

There are 577,710 inventoried and classified bridges on and off the federal-aid system. Two categories exist for deficient bridges: structurally deficient or functionally obsolete. As defined by the Federal Highway Administration (FHWA) a *structurally deficient bridge* is one that is either restricted to light vehicles only, is closed, or requires immediate rehabilitation to remain open. A *functionally obsolete bridge* is one in which the deck geometry, load capacity, clearance, or roadway alignment no longer meets the needs of the system of which it is part. Table 7-1 breaks down the number of deficient bridges on and off the federal-aid system.

Overall, 41.2% of the nation's bridges have been rated either structurally deficient or functionally obsolete. Although there has been some success in reducing the number of functionally obsolete bridges, the number of structurally deficient federal-aid bridges has jumped 32% from 1982 to 1988. The FHWA has estimated that it will take an investment of $50.7 billion to bring bridges in both categories up to current standards. Because of the age of our infrastructure, bridge needs will increase with time and will probably accelerate

**TABLE 7-1   THE NATION'S DEFICIENT BRIDGES, 1982–1988 (thousands of bridges)**

| Year | Federal-Aid | | Off-System | | Total |
| | Structurally Deficient | Functionally Obsolete | Structurally Deficient | Functionally Obsolete | |
|---|---|---|---|---|---|
| 1982 | 28 | 42 | 104 | 79 | 253 |
| 1983 | 31 | 43 | 105 | 81 | 260 |
| 1984 | 33 | 42 | 107 | 78 | 260 |
| 1985 | 35 | 40 | 100 | 68 | 243 |
| 1986 | 36 | 41 | 95 | 72 | 244 |
| 1988 | 37 | 40 | 99 | 63 | 239 |

Source: The Status of the Nation's Highways and Bridges: Conditions and Performance, Federal Highway Administration, U.S. Department of Transportation, Washington, D.C., 1989, p. V-11.

as many of the highway bridges pass the 40- to 50-year mark, which is a rough estimate of the life of many of these structures.

## BRIDGE DETERIORATION AND FAILURES

There are three basic causes of bridge deterioration and failure: poor maintenance, poor design (and or misuse), and accidents.

1. *Poor Maintenance.* Unless a bridge is properly maintained through cleaning, painting, and lubrication, a buildup of debris, dirt, and salt will ultimately damage the structure. Additionally, inspections must be performed on a regular basis to ensure that the bridge is functioning properly. Bridge supporting elements lose strength through corrosion. In steel, corrosion is caused by a mixture of dirt, salt, oxygen, and water. An electrolytic reaction begins with one part of the steel acting as an anode and another part as a cathode. This corrosion cell activity is pernicious and under certain conditions can cause a rapid loss of steel and consequently, of strength. John Fisher, a leading expert on structural steel, has argued that under especially severe conditions, as much as $\frac{1}{4}$ inch of steel as measured by web thickness can deteriorate within four years (Fisher, 1988). Reinforced concrete also suffers because the steel reinforcement bars embedded in the concrete corrode and expand, which causes spalling and cracking of the concrete.

2. *Poor Design (or Misuse of a Bridge).* There are several types of design-related failures. One is a designer's oversight of stresses caused by unforeseen loads. Such was the case with regard to the Tacoma Narrows Bridge in Washington State. Strong enough to carry traffic, it was toppled by a moderately

strong wind only months after completion in 1940. A second problem involves the use of a bridge for loads beyond its original design. Many nineteenth-century bridges could accommodate horses and buggies and subsequently cars, but could not accommodate today's tandem trucks. A bridge element may reach a point at which it is "fatigued," and suddenly yield or crack. Fatigue is most likely to be found in elements subject to frequent and significant changes in forces. Until recently, fatigue was thought to affect only steel, but engineers now recognize that it may also damage concrete bridges.

3. *Accidents.* There are numerous kinds of accidents that can damage a bridge, such as motor vehicles and trucks striking a part of the bridge and causing steel damage, or ships colliding with main supports. When an accident does cause damage, it is the bridge engineer's responsibility to learn from the problem and incorporate the lessons into future designs.

As is often the case with the country's public works, it frequently takes a disaster or an emergency before officials and the public respond. The collapse of the Mianus River Bridge in Connecticut in 1983 was caused by the failure of one of the four pin hangers supporting the span. This led to a new emphasis on "hands-on" inspections of nonredundant structural components and their elimination from future designs. Nonredundant members are those parts of a bridge that are indispensable to the structure. New York State has introduced a "fracture critical members" program which identifies nonredundant members whose failure would result in a structural collapse. Redundant steel members are being installed on many of these bridges.

Another example of a collapse that has led to a reevaluation of the bridge program is that of the Schoharie Bridge on the State Thruway in upstate New York. On the morning of April 5, 1987 two spans of the five-span Schoharie Creek Bridge collapsed, followed by a third span an hour and a half later. Five vehicles were on the bridge when it went down. Ten people died.

The New York State Thruway is the longest toll-supported superhighway in the United States. It has carried millions of vehicles since it opened in the early 1950s, and has become a vital link between New York City, Albany, and the western parts of the state. The Schoharie Bridge opened in 1954. It was a simple five-span bridge of 540 feet, the longest span being 120 feet. Problems were encountered immediately. Roadway approach slabs settled, vertical cracks appeared on the piers, and expansion joints were out of line. These problems were corrected, however, and like other bridges on the Thruway, the Schoharie was inspected periodically, and maintenance and repairs were performed by either state workers or contractors. Because of heavy rain and strong currents, before it collapsed, the soil underneath the concrete bases of the piers had washed away, a phenomenon known as *scour*.

The Schoharie was inspected in 1983 and again in 1986. In 1983 the bridge piers received a rating of 5, which meant that they were found to show only minor amounts of deterioration. The 1986 inspection was conducted in the high-water month of April. Consequently, the bottom of the piers could not be rated, and a later inspection was required.

There was reason to perform the underwater inspection with dispatch. Two years to the month before the Schoharie went down, the Chickasawobgue Bridge near Mobile, Alabama collapsed. Although no one was killed or seriously injured, the collapse prompted insurance companies nationally to require underwater inspections for similar types of bridges that had not received such inspections for at least five years. In December 1985 the New York State Thruway Authority identified the Schoharie as a prime candidate for inspection. It was planned that the inspections would begin in 1986. However, this program was delayed and the inspections were rescheduled for 1987. The bridge collapsed before the scheduled inspections took place (New York State Thruway Authority, 1987).

## BRIDGE NEEDS

*[The Brooklyn Bridge will last] as long as we are smart enough to keep the materials in the condition in which Roebling left them. But that has not been the case over the last hundred years because tax-supported structures, as compared with toll-supported ones, have not been provided with adequate maintenance funds.*

*—Blair Birdsall, 1983*

### Inspections

A comprehensive inspection program is essential for a well-managed and efficient bridge maintenance program. Inspections determine the strength and durability of a bridge and identify needed repairs. The standard biennial inspection is a visual review of every important element of a bridge. All bridge components are surveyed and given a numerical rating from 1 to 9, from which an overall rating for the bridge is reached. A special rating is assigned to elements that are inaccessible.

The standard inspection should be a first line of defense in protecting a bridge—and should identify repair needs before costly or irreversible structural damage has occurred. The inspection also highlights urgent repairs, referred to as *flags*. Should there be imminent danger to the public, the bridge inspector may have the authority to close a bridge immediately.

The biennial inspection, which is adequate for bridges in good condition, may not be sufficiently thorough and or frequent enough for bridges in poor condition. Some local governments have supplemented the federally dictated biennials with interim inspections. These inspections may be more in-depth, whereby every square inch of a bridge is thoroughly examined. Physical tests are also conducted, including the evaluation of the deck through core samples and ultrasonic testing to assess the extent of corrosion. The in-depth inspection is also used to determine rehabilitation needs.

## Maintenance

Although it may not look like one, a bridge is a machine with many moving parts. It rises and falls with loads and it expands and contracts with the weather. Measurements taken on the Manhattan Bridge in New York City have found vertical displacement as high as 8 feet with maximum load. Bridges need a great deal of maintenance to ensure that movement takes place as planned. In addition, exposed members need protection from the natural elements.

Preventive maintenance maximizes the life of the country's bridges and reduces long-term costs. The objective of a preventive maintenance program is to keep bridges that are currently in good or very good condition in those categories, rather than letting them deteriorate only to have to rebuild them later. Table 7-2 provides a breakdown of a typical preventive maintenance plan for a large bridge. A comprehensive program would consist of the following:

1. *Steel Protection.* Painting or coating is perhaps the single most important maintenance need of a steel bridge. It protects the bridge from weather conditions, air pollution, and deicing chemicals, all of which can cause structural damage over time. Bridges should be painted completely on regular cycles. Eight years is generally considered sufficient. For some bridges, such as those heavily exposed to salt, splash zone painting on sensitive areas should be completed every two or three years, with spot painting of blistered or rusted spots every four years.

2. *Dirt and Water Control.* Commonly viewed by budget directors as just for aesthetics and therefore expendable, cleaning is actually one of the most important and necessary protective actions. As described in the section on bridge failures, lack of cleaning ultimately leads to corrosion cell activity. Bridges must be cleaned on a regular cycle to prevent the buildup of dirt and salt. Additionally, drains should be regularly cleaned out so that water and other solvents are disposed of swiftly.

3. *Lubrication.* Moving parts and bridge machinery must be lubricated on a continual basis to ensure proper movement. For movable bridges, the electrical devices must be maintained and other mechanical components oiled.

# TABLE 7-2 TYPICAL PREVENTIVE MAINTENANCE PLAN FOR A LARGE BRIDGE

| Maintenance Activity | Bridge Classification | Relevant Data | Freq. per year | Total Quantity | Productivity per Crew Day | Total CD | Yearly Labor Cost | Yearly Equipment Cost | Yearly Material Cost | Total Yearly Cost |
|---|---|---|---|---|---|---|---|---|---|---|
| Debris removal | Through truss | Total length 1,420 ft | Daily | | 800 ft | 1 Crew | $173,650 | $25,050 | | $198,700 |
| | Other | 4,780 ft | Daily | | 3,000 ft | | | | | |
| Sweeping | All | 6,201 ft | 26 | 161,230 sq ft | 10,000 ft | 17 | 3,230 | 1800 | | 5,030 |
| Cleaning of drainage system | All | 6,201 ft | 1 | 6,201 ft | 200 ft | 31 | 28,750 | 5240 | | 33,990 |
| Cleaning of abutment and pier tops | All | Number of abutments and piers, 10 | 0.5 | 5 | 5 | 1 | 3,134 | 170 | | 3,304 |
| Cleaning of open grating | Bridges with open grating | Total area: 37,054 sf | 1 | 37,054 sq ft | 8,000 sq ft | 5 | 4,575 | 640 | | 5,215 |
| Cleaning of expansion joints | All | Expansion joint length, 956 ft | 1 | 956 ft | 200 ft | 5 | 7,750 | 635 | | 8,385 |
| Washing of deck and salt splash zone | Through truss | Total deck area 240,000 sf | 1 | 240,000 sq ft | 10,000 ft | 24 | 37,200 | 2976 | | 40,176 |
| | Other | 740,840 sf | 1 | 740,840 sq ft | 20,000 ft | 37 | 57,350 | 4588 | | 61,938 |
| Painting of steel preparation 25% area | Through truss | Est. steel area 433,250 sf | 0.125 | 54,160 sq ft | 13,000 sq ft | 4 | 12,556 | 692 | 1,338 | 14,566 |

| | | | | | | | | | | |
|---|---|---|---|---|---|---|---|---|---|---|
| sandblast | Other | 335,660 sf | 0.125 | 41,960 sq ft | 16,000 sq ft | 3 | 9,402 | 519 | 1,003 | 10,924 |
| 75% area SSPC-SP2 | Through truss | 1,299,700 sf | 0.125 | 182,460 sq ft | 4,500 sq ft | 41 | 128,494 | 7093 | 13,710 | 149,297 |
| | Other | 1,006,970 sf | 0.125 | 125,870 sq ft | 7,000 sq ft | 18 | 56,412 | 3114 | 6,020 | 65,546 |
| Painting | Through truss | 173,300 sf | 0.125 | 21,660 sq ft | 20,000 sq ft | 1 | 3,134 | 173 | 335 | 3,642 |
| 10% area spray | Other | 134,270 sf | 0.125 | 16,785 sq ft | 32,000 sq ft | 1 | 3,134 | 173 | 335 | 3,642 |
| 90% area brush | Through truss | 1,559,700 sf | 0.125 | 194,960 sq ft | 5,500 sq ft | 36 | 118,824 | 6228 | 12,040 | 131,092 |
| | Other | 1,208,360 sf | 0.125 | 151,045 sq ft | 8,000 sq ft | 19 | 59,546 | 3287 | 6,354 | 69,187 |
| Spot painting of steel Preparation SSPC-SP2 | Through truss | 34,660 sf | 0.25 | 86,650 sq ft | 2,200 sq ft | 40 | 57,840 | 4600 | 3450 | 65,890 |
| | Other | 268,530 sf | 0.25 | 67,130 sq ft | 3,400 sq ft | 20 | 28,920 | 2300 | 1725 | 32,945 |
| Painting 100% | Through truss | 346,600 sf | 0.25 | 86,650 sq ft | 2,800 sq ft | 31 | 44,826 | 3565 | 2,674 | 51,065 |
| area brush | Other | 268,530 sf | 0.25 | 67,130 sq ft | 4,000 sq ft | 17 | 24,582 | 1955 | 1,463 | 28,003 |
| Painting of salt splash | Through truss | 693,200 sf | 0.125 | 86,650 sq ft | 4,500 sq ft | 20 | 62,680 | 2300 | 3,160 | 68,140 |
| zone | Other | 537,050 sf | 0.125 | 67,130 sq ft | 7,000 sq ft | 10 | 31,340 | 150 | 1,580 | 34,070 |
| Preparation SSPC-SP2 Painting 100% | Through truss | 693,200 sf | 0.125 | 86,650 sq ft | 5,500 sq ft | 16 | 50,144 | 1840 | 2,525 | 54,509 |
| area brush | Other | 437,050 sf | 0.125 | 67,130 sq ft | 8,000 sq ft | 9 | 28,206 | 1035 | 1,420 | 30,661 |
| Patching of sidewalks | | Sidewalk area, 273,000 sf | 0.25 | 68,210 sq ft | 1,000 sq ft | 69 | 64,170 | 8280 | | 72,450 |
| Crack sealing in pavement and curbline | | Pavement area, 310,000 sf | 0.50 | 155,000 sq ft | 10,000 sq ft | 16 | 23,072 | 1456 | | 24,528 |
| | | | | | | 492 | $1,116,801 | $90,859 | $59,135 | $1,266,895 |

Sources: NYC Department of Transportation, Columbia University, City College of New York, Cooper Union, Polytechnic University, and Pratt Institute.
sf = square feet.

4. *Roadway Surface Maintenance.* The pavement and curbline of a bridge roadway should be properly sealed and the sidewalks should be patched as necessary to keep water from damaging the supporting members.

A rough guide to establishing the magnitude of maintenance is to set aside funds equal to a percentage of the replacement costs of bridges. A 1981 study conducted by the Organization for Economic Cooperation and Development (OECD) found a range between 0.3 and 2.5% of the replacement cost as the annual preventive maintenance needs. Japan, Germany, and Italy were the leading countries in providing preventive maintenance.

## Repair and Rehabilitation

No matter how well maintained a bridge is, some elements will need to be repaired because of accidents or because they have reached the end of their useful life. Bridge decks wear out and need to be replaced after 30 to 50 years. Large cables on suspension bridges may have lives in excess of 100 years, while diagonal cables on "stayed" bridges may last only 10 to 30 years. A well-balanced bridge program should identify the advantages and disadvantages of both low-cost repairs and more costly rehabilitation and replacement. However, most bridge programs in the United States are not driven by the most efficient methods. Restrictions placed on various pools of funds often force the engineer to choose more expensive schemes.

With proper maintenance, most of the country's steel bridges would last for centuries. That many of them have fallen into a state of disrepair is in most cases, less a comment on the original design of the bridge than on the policy that allows our public works to deteriorate through neglect. This "disposable bridges" mentality can be attributed to several causes, most notably to the way that bridge work is funded in the United States.

A variety of funds are distributed by federal, state, and local governments. However, most of the federal, state, and local bond funds are slotted for capital work (i.e., long-term rehabilitation and reconstruction) rather than expense work, which would include general maintenance activities. In the long run this policy is costly and inefficient. From a local government's point of view it may appear cheaper in the short run to let a bridge deteriorate and become eligible for federal capital funds than it is to maintain the structure properly and keep it from falling into poor condition. Figure 7-6 shows how a reduction in maintenance will actually increase the amount of capital money needed later to rehabilitate a bridge.

Oddly enough, the long-term solution to the country's bridge crisis lies not in increasing the number of dollars spent but in how we spend the dollars we have. If we achieve a steady state of all bridges in good condition, we could maintain

# LONG TERM ANNUAL COST OF BRIDGE PROGRAM

## BY MAINTENANCE LEVEL

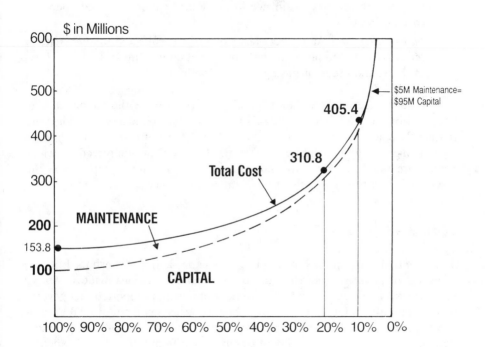

**NYC DEPARTMENT OF TRANSPORTATION**   **COOPER UNION**
**COLUMBIA UNIVERSITY**   **POLYTECHNIC UNIVERSITY**
**CITY COLLEGE OF NEW YORK**   **PRATT INSTITUTE**

**Figure 7-6**   Long-term annual cost of bridge program.

our structures for billions of dollars less than we spend today. There are various methods available for dedicating the revenue needed to rehabilitate bridges so that engineers with long-term vision, not budget directors with annual constraints, plan for the future of our public works.

1. *Direct User Fees.* A user fee would place the burden of funding a bridge's upkeep on the people who use the bridge. Tolls are the most common form of direct fee. They are usually collected by public authorities, which then use the money to finance inspection, maintenance, and operating costs.
2. *Indirect User Tax.* An indirect tax is one that is collected through state or federally imposed taxes. It could be levied on gasoline or be in the form of registration fees, licensing fees, or traffic fines.

The advantage of a dedicated funding source to meet the nation's bridge needs is obvious. Competing with other social ills for public money, bridges are frequently placed farther down on the wish list. If we invested in our bridges now, money would actually be freed in the long run for other needs, such as health care and schooling or the reduction of taxes. Figure 7-7 depicts how the service life of a bridge is a function of the maintenance level.

## CONCLUSION

The experience of the Silver, Mianus, Schoharie, and Williamsburg Bridges, along with others, substantiates the major recommendations made in *Fragile Foundations.* The Council on Public Works Improvement reported that government expenditures (at all levels) for public works as a percent of total government spending has dropped from 19.1% in 1950 to 6.8% in 1984. In the same period, interest on the general debt has increased from 8.6% of total expenditures to 10.1%.

But money alone is not the answer. A recurring theme throughout the report is the need to increase maintenance. The council specifically recommended "that budgetary priority go to the funding of maintenance of existing facilities." Current practices frequently restrict federal money to capital work and contribute to the disposable mentality that has affected so much of American life.

The nation is at a decisive point. At the beginning of the twentieth century and again following World War II, the United States entered into a period of extensive public works construction and became the envy of the world. Today, the country needs to reinvest wisely in its public works to guarantee that future generations have access to these facilities.

# REDUCED MAINTENANCE

## BRIDGE SERVICE LIFE AS A FUNCTION OF MAINTENANCE LEVEL

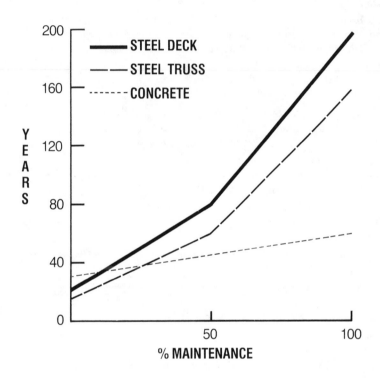

Reference: **BRIDGE MANAGEMENT SYSTEMS** (Draft)
Transportation Research Board, NRC

NYC DEPARTMENT OF TRANSPORTATION        COOPER UNION
COLUMBIA UNIVERSITY                                        POLYTECHNIC UNIVERSITY
CITY COLLEGE OF NEW YORK                              PRATT INSTITUTE

**Figure 7-7** Reduced maintenance: bridge service life as a function of maintenance level.

## ADDITIONAL READING

Birdsall, Blair, The Brooklyn Bridge at 100, *Technology Review*, April, 1983.

Cather, Willa, *Alexander's Bridge*, New York, 1987 [1912].

Federal Highway Administration. *The Status of the Nation's Highways and Bridges: Conditions and Performance* and *Highway Bridge Replacement and Rehabilitation Program*, Washington D.C., 1989.

Hindle, Brooke, Spatial Thinking in the Bridge Era: John Augustus Roebling versus John Adolphus Etzler, in *Bridge to the Future: A Centennial Celebration of the Brooklyn Bridge*, Margaret Latimer, Brooke Hindle, Melvin Kranzberg, eds. New York, 1984.

Kemp, Emory L, Roebling, Ellet, and the Wire-Suspension Bridge, in *Bridge to the Future: A Centennial Celebration of the Brooklyn Bridge*, Margaret Latimer, Brooke Hindle, Melvin Kranzberg, eds. New York, 1984.

Mair, George, *Bridge Down: A True Story*, New York, 1982.

Martin, Albro, 'Downtown': The American City in the Railroad Age, in *Bridge to the Future: A Centennial Celebration of the Brooklyn Bridge*, Margaret Latimer, Brooke Hindle, Melvin Kransberg, eds. New York, 1984.

National Council on Public Works Improvement. *Fragile Foundations: A Report on America's Public Works*, Washington D.C., 1988.

National Transportation Safety Board. *Highway Accident Reports: Collapse of New York Thruway (I-90) Bridge Over the Schoharie Creek, Near Amsterdam, New York, April 5, 1987*, Washington D.C., 1987.

New York City Department of Transportation. *Spanning the 21st Century: Reconstructing a World Class Bridge Program*, New York, 1988.

New York State Thruway Authority. *Collapse of the Thruway Bridge at Schoharie Creek*, Albany, 1987.

Organization for Economic Cooperation and Development. *Bridge Maintenance*, Paris, 1981.

Patton, Phil, *Open Road: A Celebration of the American Highway*, New York, 1986.

Reier, Sharon, *The Bridges of New York*, New York, 1977.

Steinman, David. *The Place of the Engineer in Civilization*, Raleigh, N.C. 1939.

Vose, George, *Bridge Disasters in America: The Cause and the Remedy*, Boston, 1887.

## QUESTIONS

**1.** Describe the two periods of bridge construction in the United States.

**2.** What explains the increase in highway and bridge construction following the second World War?

**3.** Describe the six major bridge types and explain the characteristics that they share and what makes each distinct.

4. Describe at least two innovations used by John Roebling in his design for the Brooklyn Bridge.

5. What bridge disaster led to a national program to inspect the nation's bridges?

6. What are the federal regulations regarding bridge inspections?

7. What is the difference between a structurally deficient bridge and a functionally obsolete one?

8. What are the three major causes of bridge deterioration and failure?

9. Describe the four major components of a comprehensive maintenance program.

10. What is the argument behind the statement: ". . . the long-term solution to the country's bridge crisis lies not in increasing the number of dollars spent but in how we spend the dollars we have."

Work Barges, East River Drive, New York

# 8

# WATERFRONT INFRASTRUCTURE

**Ray Gordon**
Office of Mark A. Kates, Architect

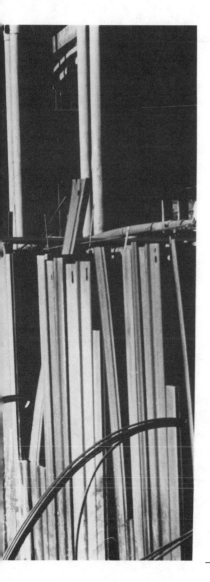

## THE WATERFRONT ENVIRONMENT

The design and construction of the infrastructure serving waterfront sites present the professional engineer and architect with a number of unique challenges. Wave action, salt spray, high winds, erosion, storms, and variable subsurface conditions combine to make the waterfront environment quite different from the conditions typically found on upland sites.

The most significant difference between building structures on inland parcels and construction along the water's edge is the projected useful life of each type of structure. Given proper design standards and periodic maintenance, the average building constructed on inland sites can be expected to last virtually indefinitely. However, because of the wide variety of adverse conditions associated with waterfront sites, shorefront structures have a finite life that can be extended significantly only by allocating substantial amounts of money and resources in the rehabilitation and reconstruction of key structural components. Thus construction techniques that are perfectly acceptable for inland sites become problematic when applied to most waterfront sites. In many cases, structural systems on the waterfront are exposed directly to potential damage from wind, water, erosion, and impact from waterborne craft. The designer must select materials carefully not only for their structural properties but also for durability in an often hostile natural environment.

### The Water's Edge

There are four basic ways in which bodies of water meet the land: beaches, marshland, man-made barriers, and exposed rock formations. Beaches are found along ocean coasts and along some lake and river shores. Composed of fine particles of eroded rock, beaches create a gradual transition from land areas to deep water. A beach's sand particles are subject to significant movement in a relatively short period of time. The shifting nature of a beachfront makes difficult accurate definition of the land area immediately adjacent to the water and can cause significant structural problems in the construction of foundations and the placement of utility lines.

Marshes can be either freshwater wetlands—near ponds, lakes, streams, and some rivers—or saltwater wetlands located along rivers, bays, and barrier islands near the ocean. If the marsh is subject to tidal action, the wetlands plants may be partially or totally submerged at high tide, and completely exposed on relatively dry land at low tide. Subsurface conditions are often poor, with saturated soils and layers of organic material which can move laterally when vertical loads are applied. Wetlands are also an important habitat for a multitude of plants, animals, and migratory birds. For this reason, most municipalities and state governments have placed restrictions on the use of lands in and near wetlands.

In areas where man-made structures have been placed, the edges are usually defined by structures designed to retain the earth behind the barriers and protect the upland areas from erosion on the water side. Bulkhead, riprap walls, and similar structures are also used when land areas have been extended by means of filling areas that once were under water or below the grade of adjoining properties.

Natural rock formations can be found along every type of water body. Occasionally, beaches have been formed along the water in front of rock formations. Rock outcrops can range from small ledges to tall cliffs formed by erosion over many millennia. Access to these shorelines is often complicated by steep slopes and significant grade changes between the upland areas and the water's edge.

## Tides and Currents

The water levels of oceans and their associated bays and inlets are subject to changes caused by the gravitational pull of the moon and sun on these large bodies of water. One tidal cycle takes approximately 12 hours to complete, with two high tides and two low tides in each 24-hour period. Tide elevations are generally measured from mean sea level or a datum point using sea level as a point of reference. The range in elevation from mean high tide to mean low tide can be substantial, and varies from location to location. Generally, for design purposes the most important figure is mean high water, which is the average height of high water recorded over previous years at a specific location. Because the change in water levels can average several feet or, in extreme cases, scores of feet, the elevations of both high and low water is an important design factor, especially when a project incorporates a floating element such as a dock or vessel.

Currents are present in bodies of water that are subject to movement caused by temperature differential, tidal action, or gravity-fed streams and rivers. Generally, given a constant depth of water, the velocity of a current will increase if the banks of the river or inlet are narrow and constrict the flow of water parallel to the banks. Conversely, if portions of the water body widen, the current will slow down as the volume of water is spread across a greater horizontal distance. Ocean inlets and river mouths opening onto ocean bays are subject to the effects of tides on the flow of water. As the tide rises, water flows into inlets, bays, and adjoining rivers, creating a current moving upriver away from the ocean. As the tide falls, this water returns to the ocean and the current reverses itself.

Rivers and streams are usually fed by runoff from their watersheds at higher elevations. The volume of water at any given time is therefore dependent on the amount of surface water present in the watershed. After heavy rains, the volume of water increases, and the natural currents pick up speed. Droughts cause a reduction in the volume of water, while snow melts increase runoff. Seasonal changes are most evident in temperate climates and in locations subject to regular rainy or dry seasons. The most important factor to consider when designing along a stream or riverfront is the potential for extreme storms, which can cause localized flooding conditions and create very strong currents that may erode the riverbanks and damage flood protection structures. In coves or portions of a river or inlet that have unusually shallow or deep areas, currents can form eddies that swirl within a confined area and can create unpredictable currents and whirlpools. Over time, these conditions can accelerate erosion along the shoreline.

## Wind Conditions

Because of the exposure created by a large body of water, prevailing winds along a shorefront are more noticeable and usually have greater force than do winds at inland sites. Winds also have a direct effect on the growth patterns of vegetation, with trees bending away from the constant pressure of winds blowing off the water. Some trees and shrubs, while otherwise well suited for the local climate, cannot tolerate the high winds at waterfront locations.

Storm conditions are more pronounced at the waterfront, and winds can increase water levels, wave heights, and the velocity of rain and snow. Wind forces must be considered in the design of upland buildings, shore protection systems, and in-water structures. Strong winds can exert large horizontal loads against vertical solid masses, which behave like sails in high winds. In the case of an immovable structure, these forces can cause significant damage. Bulkheads and retaining walls are subject to repeated strikes by wind-driven waves and salt spray. Vessels with superstructures also catch the wind and can be driven against docks and bulkheads with substantial force in even moderate winds if not properly secured.

## Subsurface Conditions

Subsurface conditions along a waterfront can vary considerably. Bedrock is most often close to the surface along riverfronts and landlocked bodies of water. Along ocean shorelines, bedrock is usually located beneath layers of sand and silt. Soils range from sand and gravels, which are generally well drained, to less well drained soils such as silts, clays, and organic materials. Both natural and man-made forces contribute to a constantly changing waterfront topography. Many waterfront properties have been built on land that was once below water level. This land could have been created by natural siltation, through storm, action, or by active landfilling.

Waterfronts historically have been used extensively for landfilling operations. In fact, in many waterfront cities, one of the most prevalent means of enlarging the municipal land area was originally to place landfill in the water adjacent to the existing upland. This process has continued in some locations, although increased regulation has limited landfilling operations in most jurisdictions.

Landfills present two major problems: unconsolidated fill and the potential for toxic materials. Landfilling often results in voids or poorly compacted areas. These areas are very difficult to locate and usually are not detectable from ground level. Materials placed in landfills often contain industrial waste, machinery, lubricating oils, heavy metals, or similar compounds which are toxic in significant concentrations. Because of the wide variety of subsurface conditions, borings and soil testing are absolute necessities in the early stages of project development. If the tests reveal the presence of hazardous materials, the removal of these materials should be arranged prior to beginning construction.

# FIXED STRUCTURES AND THEIR USES

There are several types of waterfront structures. Their uses include commerce and industry, shipping, erosion protection, and transportation.

## Piers and Wharves

Piers and wharves are structures supported by a grid of bearing piles. The piles are usually friction piles but can be driven to bedrock to support heavy loads. Piers are situated at an angle to the shoreline and are typically constructed perpendicular to a bulkhead or retaining wall. Wharves are similar to piers in construction and function, the primary difference being the orientation of the structures parallel to the shoreline. Piers were originally constructed to serve the shipping industry as platforms to load and unload cargo vessels. Piersheds were constructed over the piers to enclose cargo while it is awaiting transfer onto ships or onto carriages and trucks for transport to market. Piers were also used as areas for recreational fishing in shallow water.

With the coming of containerized shipping, traditional piers and wharves were no longer able to serve the shipping industry efficiently. Piers with piersheds were converted for use as storage facilities, while open piers were utilized to park vehicles or store construction materials, or were left underutilized. Wharves were occasionally constructed at containerports to serve as loading platforms for materials lifted from ships on the huge container cranes. Wharves also continued to be used for commercial fishing operations, as an efficient staging area for unloading fish, repairing fishing equipment, and servicing trawlers. These wharves often had large storage and processing sheds nearby, constructed parallel to the water's edge.

Currently, there is a revival of public interest in the waterfront. This has encouraged a number of new uses for underutilized or abandoned pier structures in many municipalities, including restaurants, concert facilities, tennis clubs, shopping malls, maritime museums, amusement parks, exhibition spaces, and residential developments.

## Platforms

Platforms are structures built on piles designed to support large-scale projects. Platforms are larger than piers and are often designed specifically for commercial or industrial use. Platforms are sometimes constructed as relieving structures to buttress adjoining retaining walls while providing additional space for public use or commercial activity. Platforms are often constructed as an alternative to landfilling. Platforms of several acres can accommodate multistory residential buildings, office buildings, public parks, or shopping facilities while appearing to be extensions of the adjoining land area.

## Bulkheads and Retaining Walls

Bulkheads are vertical structures placed at the edge of a land area to retain the earth while protecting the landmass from erosion from the water on the opposite side. The structure can be placed upland from the water's edge or in the water itself. Bulkheads can be constructed of wood sheathing, steel sheet piling, or concrete. Piles are usually placed on the face of the bulkhead to support the sheathing. Whalers are placed between the piles for stability. For added strength, bulkheads can be supported by steel cable tiebacks to concrete deadmen placed upland from the bulkhead. The deadmen are often pile-supported to maintain their position and to prevent excessive settling.

Riprap walls are retaining walls constructed of large rocks or pieces of concrete. The rock or concrete blocks are placed at an angle (approximately 45° to the ground plane) and interlock to form a strong wall. The riprap extends several feet below low water, anchored with large pieces of rock at the base of the wall to prevent slippage of rock into the water. The area immediately behind the wall is backfilled with gravel, sand, or earth and is compacted. The area next to the top of the wall is usually topped with a paved area to minimize the infiltration of water behind the wall. Screening can be placed behind the wall to help limit the erosion of earth between the rocks. There are also manufactured concrete systems that interlock more systematically and form a more uniform wall.

Another structure used to create waterfront walls is a series of gabions. The gabions are typically constructed by stacking a group of large wire screen containers on top of each other, and filling them with rock and smaller stones. Cement is sometimes added to increase the strength of gabions, especially when the wire mesh eventually erodes away from the rock core. The gabions are placed next to each other at a slight angle away from the water toward the upland side. Gabions are more cost-efficient than bulkheads and are more stable than riprap walls.

## Groins and Breakwaters

Groins are narrow structures built perpendicular to the coastline to protect beachfronts from erosion due to wave and tidal action. Groins, or jetties, are constructed of wood, large concrete blocks, or rock, and can be used as fishing platforms or pedestrian areas. Groins protect the sand beaches on the side away from the prevailing currents, and actually encourage sand to build up on that side. However, groins cause the currents to erode the sand on the side facing the current. Thus although the groins dissipate wave energy directed to the shore, they result in very little actual gain in total area of sand on the beaches they protect. Thus the installation of groins is a controversial and costly method of erosion protection.

Breakwaters are placed in the water to protect harbors, mooring areas, and piers from damage caused by wave action, ice flows, and storm winds. They can be constructed as permanent structures at the mouths of natural harbors, as groins parallel to the shore, or as floating structures anchored beyond moored vessels. Fixed breakwaters can be built of stone, concrete, wood bulkheads, or steel piling, and work best if placed perpendicular to the direction of approaching waves.

Floating breakwaters can be built of wood or modular concrete boxes with foam cores. The floating structures work best if they are at least 10 to 15 feet wide. This width allows the floating breakwater to interrupt the wave pattern instead of simply floating over the wave as it passes undisturbed below the floats. Floating structures can either be held in place with pin piles, or anchored with a number of heavy anchors to the bottom of the body of water.

## Bridges and Causeways

Bridges can be constructed in several different ways. The most common is as a pile-supported roadbed with relatively small clear spans. Larger spans usually are built on large concrete or stone piers and include either a suspension system or an overhead arch to carry the center span. In suspension bridges, tall towers are placed above the piers and heavy abutments are necessary to secure cables on each end of the bridge. The piers are usually supported on pile foundations, unless bedrock is relatively close to the riverbed (see Chapter 7). Causeways can be constructed on narrow portions of filled land, small natural islands, man-made islands, or pile systems. Causeways are generally used to span long distances over relatively shallow waters where no clearance is required for shipping.

## FIXED-STRUCTURE DEVELOPMENT ISSUES

The following are some of the key issues that should be addressed in the planning and design of fixed waterfront structures.

### Bearing Pile Construction

Bearing piles can be constructed of wood, steel, or concrete. Generally, wood piles are the least expensive, weakest, and least durable; concrete is the most expensive and most durable. Steel piles are the strongest pile option.

Wood bearing piles can be used most efficiently in situations where a short pile is required and where very heavy loading is not anticipated. Wood piles can be fabricated from a variety of woods and densities and typically range from 30 to 60 feet in length. After being driven into the ground, the tops can easily be trimmed to a uniform elevation to accept a structural cap. Wood piles are usually friction piles which obtain their bearing strength by the friction of the compacted earth against the sides of the pile along its length. Wood piles are also used as fender piles, placed on the front of fixed structures to absorb impact loading from vessels without damaging the structure itself. When used as fender piles, the piles are not driven deep into the ground, to allow for movement in the pile and easy removal for replacement.

Steel piles are used where strength is very important. Steel piles resemble standard steel wide-flange sections but are constructed with extra-heavy webs in order to withstand the force of pile driving and vertical loading. Steel piles can easily be extended to any length and are necessary when driving to deep bedrock formations. Steel piles are usually capped with concrete to minimize the corrosive effects of salt spray and oxidation in the splash zone. Steel piles can also be constructed as tapered hollow cylinders fitted with a steel cone tip. These hollow piles can be filled with reinforced concrete for added strength and lateral stability.

Reinforced concrete piles can be constructed in square or cylindrical sections of varying length. Most concrete piles are precast and are usually used in applications where the depth of piles is known with some certainty. They are often fitted with a steel tip to assist in pile driving. The concrete piles are excellent in resistance to corrosion, are very strong, and are often used in conjunction with precast concrete structural systems and pier decking. Concrete piles are occasionally cast in place within holes drilled in the ground. The ends are also "bulbed" to resist uplift forces.

## Transportation Access

The proximity of a waterfront site to adequate access roads is a key development issue. Roads should be able to accommodate construction equipment, service and emergency vehicles, trucks, and automobiles. Roadbeds must be designed to resist settling due to poor subsurface conditions, and should not have an adverse effect on wetlands or areas prone to periodic flooding. This can be achieved by minimizing the amount of fill used in building causeways, providing drainage culverts below the roadbed, directing runoff into sewer systems, and routing roads around environmentally sensitive areas.

Because of the relatively flat topography and ready availability of land, railroads, major highways, and wharves are often located along the shoreline. These structures can create substantial physical barriers between the waterfront and upland access points. If such barriers should exist at a waterfront development site, at-grade crossings must be carefully planned to allow safe access to the waterfront without creating conflicts with the existing transportation system in the area. Elevated crossings should be considered only if at-grade crossings cannot be provided.

## Public Accessibility

Public access to the waterfront has become a requirement of waterfront projects in a number of municipalities and states. Whenever feasible, designers should consider ways of incorporating public esplanades, sitting areas, and overlooks into their projects. Public access brings activity to the waterfront, which is beneficial to retail or entertainment establishments. This activity must be balanced, however, with concerns for security, privacy in residential projects, and noise generation. Public spaces should also be designed without blind spots, sharp turns, or obstructed views of the water. Whenever possible, esplanades and public spaces should be linked to existing sidewalks, parks, and other public areas.

## Proximity to Existing Utilities

Because waterfront sites are usually the last areas in a municipality to be developed, basic utilities are often at a great distance from the parcels. The availability of utilities should therefore be one of the first items researched before beginning a waterfront project. In many cases, sanitary and storm sewers do not adequately serve low-lying waterfront parcels. The low waterfront site elevations, coupled with existing sewers that are relatively close to surface, makes the construction of gravity lines able to reach distant sewer mains problematic. This usually requires the construction of pumping stations and force mains to pump storm water and sewage to remote connections into the existing gravity sewers. In areas where the subsurface is not stable, sewers must be supported by piles, adding considerable expense to the infrastructure construction budget.

Although the construction of water, electric, telephone, and natural gas lines poses fewer logistical problems, there are several items to be considered. Landfill sites may contain large pieces of debris, which must be removed prior to the completion of utility lines. The settling of the land may twist subsurface utility conduit and chases, which may result in breakage and leaks. Also, the added distance of waterfront utility runs are significantly more costly than the installation of utilities to serve typical inland sites.

## Groundwater

The water table is usually very close to the surface at waterfront locations. Near oceans, bays, and inlets, the water table may consist of salt water or a brackish combination of salt water infiltrating the fresh groundwater. The high water table presents constraints on construction and limits the depths of excavation. Basements and foundation walls should be protected against water pressure acting on below-grade structures, and footing piles are usually required to reach adequate bearing surfaces. Freshwater aquifers may also be located beneath a waterfront site. These aquifers may be sources of potable water supplies for municipalities, and measures must be taken to protect aquifers from spills from underground fuel tanks and runoff to recharge basins.

# VESSELS AND FLOATING STRUCTURES

Vessels of all types are found on waterways, including cargo vessels, pleasure craft, tugboats, maintenance craft, and houseboats. In addition, there are a growing number of floating structures used for retail, food service, and recreation purposes. The following is a brief description of the principal types of waterborne vessels and structures that may be found along a waterfront.

## Cargo Vessels

There are five basic types of cargo vessels: break-bulk carriers, container ships, tankers, railcar ships, and bulk cargo ships. Break-bulk carriers were once the predominant type of commercial ship and dominated shipping until the advent of containerized cargo handling in the 1960s. "Break-bulk" refers to individual items of cargo, such as sacks of grain, machinery parts, barrels, and boxed goods, which are gathered in nets or loaded on palettes and placed in the holds of a cargo ship. When loading or unloading such a vessel, cranes lift each piece of cargo and move the item between the ship's hold and the pier deck. This is a time-consuming and labor-intensive method of handling cargo. A typical ocean-going ship would take an average of 2 to 3 days to completely load or unload and would require large crews to load the material onto the crane lift, remove the material from the lifts, arrange the cargo on the pier in organized stacks, and then load the cargo onto waiting trucks.

Container technology radically changed the shipping industry. Instead of shipping small palettes of cargo, goods were loaded into large containers, up to 8 feet by 8 feet and 40 feet long, at the point of departure. When arriving at its destination, the container is lifted directly off the ship and either into a storage area or onto a waiting trailer truck for delivery to its destination. This process streamlined the handling of cargo, and ships could be unloaded in less than a day with a fraction of the labor hours previously needed to accomplish the task. This ease of cargo handling also allowed cargo ships to be larger and to serve multiple ports in one voyage. These ships can be up to more than 800 feet in length and can carry up to 4000 containers.

Container ports require significant upland storage areas, with the average five-crane facility having upward of 120 acres of staging areas adjoining the loading area. The cranes used to load container ships are multimillion-dollar mobile structures with horizontal booms that reach over the ship and lift the containers vertically off the ship and down onto the adjoining storage areas or trucks in one operation. Mobile cranes are placed on rails parallel to the ship berths, and the substructure must be constructed on piles that can withstand considerable loads.

Tankers come in a wide range of sizes, from huge crude oil tankers over 1000 feet long to small 100-foot lighters used to off-load oil from larger ships for delivery to individual port tank farms or waiting ships. Tankers are typically used to transport oil, liquid natural gas, fuel, or other liquids shipped in large quantities over long distances.

Ships are also used to transport railroad cars from one port to another without unloading the railroad cars themselves. These ships are often called "ro-ros," an abbreviation of "roll-on and roll-off." The cars are rolled onto a ramp at the stern of the ship onto rails on the decks of the ship. When reaching their destination, the ramp is lowered and the cars are rolled off the ship and into the railyard at the port. Bulk carriers are ships with large interior holds or compartments designed to hold loose materials. Typical cargoes include grain, coal, gravel, or other materials that can be loaded by hoppers or pipes directly into the holds.

## Barges

Barges are the most versatile vessels for moving material within a port facility or on protected waterways and inland river routes. Barges are usually used for cargo handling where open flattop vessels can accommodate lumber, heavy machinery, railroad cars, and other large items that are difficult to move over land. Barges with sidewalls or totally enclosed superstructures can be used to move bulk cargo, building materials that must be protected from the weather, refuse for transfer to disposal sites, or discarded lumber for burning at sea.

Barges are utilized for the storage of construction materials for repairs to bulkheads, platform construction, bridge foundation repair, and pier reconstruction work. Their significant carrying capacities allow for the delivery of large, bulky materials such as piles, which are often inaccessible from the upland, directly to waterfront sites. Barges can also be fitted with cranes and pile drivers for reconstruction and demolition work.

There have been several conversions of barges into commercial facilities such as floating restaurants. These facilities utilize either surplus barges with superstructures, or new vessels built to meet the specific needs of the commercial establishment. These "floating" restaurants and entertainment barges are usually placed atop pile-supported cribs located at approximately the low-water elevation. The barges are floated onto the crib and ballast is added to the hold to settle the barge securely on top of the crib. This provides a secure foundation that does not allow the barge to move when waves or the wake from a passing vessel hits the structure.

## Tugboats

Tugboats are the workhorses of the harbor. They assist all large ships in berthing, direct the passage of cargo ships through narrow channels, propel barges, and can be used to fight oceanfront fires. These relatively small vessels, typically between 60 and 120 feet in length, have a low center of gravity and are very powerful for their size. Working in teams, they can turn a huge tanker or cargo ship around in minutes. Because tugs are in use for many hours at one time, they require temporary docking facilities at several locations in the harbor to provide water, fuel, and supplies.

## Sailing Vessels

Sailing vessels are usually used for recreation, although there has been a recent trend toward commercial uses of large sailing ships which utilize innovative sailmaking technologies. Sail-powered vessels are less maneuverable than powered vessels and move much more slowly. Sailboats typically range from 25 feet up to ships 125 feet or more in overall length. Recreational vessels can be moored in open water in designated anchorages or at docks in marinas. Mooring areas must be kept clear of shipping lanes and of areas with strong currents or locations with unprotected storm exposure. Commercial sailing vessels are used as either chartered party boats or as break-bulk carriers. These vessels require docking facilities similar to those used by conventionally powered break-bulk ships.

## Power Pleasure Craft

Pleasure craft can be powered by outboard motors, inboard–outboard motors, or sophisticated high-powered performance engines. These vessels can range from tiny flat-bottom fishing boats only a few feet long to luxury yachts over 300 feet long. The smaller craft should be located with the same precautions as are used in siting sailboat moorings, while the larger vessels are usually treated as sailing ships which require docking facilities much larger than those required by the average pleasure craft. For the year-round berthing of pleasure craft, full-service marinas are required which provide repair facilities, dry docks, slips of varying sizes, basic utilities and stations for sanitary waste pump-out.

## Houseboats

Some vessels are designed specifically for use as living quarters. These boats are usually under 50 feet in length and are not designed for extensive travel or rough-weather conditions. Houseboats are moored in marinas or at docks that provide basic utilities throughout the year. Security is a key issue with vessels designed to have people living aboard, especially in highly populated urban areas.

## Marinas

Marinas should provide water, electrical, telephone, and waste disposal facilities for all vessels. They usually also have ancillary facilities such as a dry dock, repair shop, chandlery, restaurant, showers, equipment sales and rentals, boat sales offices, and upland areas for the winter storage and refurbishing of boats. Some marinas are no more than boat basins for the mooring of boats without support facilities or upland storage areas. All marinas and boat basins should be protected from storms by means of breakwaters, jetties, or narrow harbor mouths.

## FLOATING-STRUCTURE DEVELOPMENT ISSUES

The following are several of the issues that must be addressed when designing waterfront areas to accommodate floating structures and vessels.

### Stability

For a vessel or floating structure to be used properly, it must be stable when placed in the water. The structure should have a low center of rotation, which places the majority of the mass of the structure as low as possible. In general, the lower the center about which the mass can be rotated, the more stable the structure and the more resistant to movement caused by shifting loads, wind, waves, and vessel wake.

The relationship of the mass of the structure to its flotation capacity is also an important consideration. There is a tendency to provide the greatest amount of flotation capacity; however, if the structure is large and lightweight, it may be too easy to move. This is especially evident in docking systems, which often are made of lightweight plastics or foam-filled shells. Although adequate for light-duty applications, such structures would not be stable if used by large numbers of people or if subject to impact loading. When designing docks and vessels to accommodate large numbers of people, a more massive structure with a low center of rotation would be preferable, as long as there is sufficient flotation to exceed the combined design dead and live loads.

The relationship of the height of the structure to the width of the base in the water is also important. If the superstructure is too tall, not only does the center of rotation rise, but the superstructure itself can act as a sail in high winds. A structure with a wider base is generally more stable than a narrow vessel, especially when oriented parallel to the wind or prevailing currents.

Cradles can prevent the movement of a structure, which is a significant problem for restaurants and public assembly facilities. Cradles can be constructed below the high-water mark to provide a solid base for the placement of structures in the water. A series of piles is placed in the water, and a grillage, or cradle, is constructed above in the approximate dimensions of the vessel to be secured. The barge or structure is then floated in place above the cradle, ballast is added, and the structure settles on top of the pile-supported structure. If repairs are needed, the ballast can be removed and the structure floated to dry dock.

Floating structures can also be held in place using pin piles. Pin piles are driven through hoops attached to the sides of the floating structure. This allows the structure to rise and fall with the tide and over waves without much lateral movement. Pin piles are used most often to secure floating docks and walkways in marinas.

## Freeboard Height

The height of the top of the deck of a floating structure should be carefully matched to the dock height. The differential between the heights should not be so great as to cause access ramps to have too steep an angle at low tide. The tidal range should also be taken into consideration so that the floating structure's deck is not significantly above the dock or pier deck at high tide.

## Load Capacities

The ability of a floating structure to accept adequate live loads is a crucial design issue. Floating docks should be able to hold at least 50 pounds per square foot before the deck approaches the water level. Vessels used to carry passengers and cargo should have substantially greater live-load capacities. Normally, pier decks are designed to accept live loads in excess of 150 to 200 pounds per square foot. Piers should always be designed to hold emergency vehicles and maintenance trucks which will be required for periodic maintenance of the structure.

## Utility Connections

Utility connections between upland sources and floating structures must be flexible. Electrical lines should be of approved marine double-insulated cable. If possible, avoid placing electric lines in rigid metal conduit at points where torsion in the line may occur. Adequate slack should be provided so that the electric lines can reach through the entire range of tides, from low low water to high high water. Water lines must have special fittings and flexible connecting hoses from the fixed upland connection to the mains. Waste lines can be provided only if pumping stations are provided in conjunction with flexible pipes. Alternatively, waste can be collected on the structure itself and pumped out periodically to service vessels. Telephone lines usually can be connected directly as long as adequate slack is provided in the line.

Utilities should also be resistant to corrosion. Although marine-grade aluminum can be used, metallic conduit should be replaced with nonmetallic materials when feasible. Utility lines and conduits should also be kept away from direct contact with salt spray in locations such as pier decks or along bulkheads.

When providing electrical service to vessels, one should keep in mind that several types of service are required by different boats, depending on their size and power requirements. Utility connections should allow for a number of electrical phasing options. This can be accomplished by installing utility connection devices which provide differing service based on the type of specialized connector that is attached to a universal outlet. This simplifies the provision of electric service in most cases where vessels are transient or are berthed for a single season.

## Berthing Conditions

Several berthing systems are available for floating structures and vessels. They can be berthed parallel to a pier, with mooring lines fore and aft and diagonal spring lines to minimize lateral movement. More than one vessel can be moored alongside each other using a series of similar mooring lines. Boats can also be berthed stern-first against a bulkhead or pier. In this case, the boat is usually tied to at least one set of piles perpendicular to the bulkhead, while additional lines may be secured to piles along the bulkhead itself. This method of berthing is useful if the boat has a relatively low freeboard and has provisions for loading passengers from the stern.

Pleasure craft berthed at floating docks are usually tied to a row of piles on either side of the berth, not to the floats themselves. Craft are also often moored to anchors away from docks or piers. These designated anchorage areas provide a relatively secure location accessible from the dock only by launch. These anchorages are the safest locations for a boat to ride out a storm, since there is no nearby dock against which the vessel may come into contact with significant force.

## Fendering and Vessel Protection

There are a number of methods employed to keep vessels or floating structures from significant damage while berthed against a fixed structure. The most common is a series of fender piles, which are essentially sacrificial wood piles secured loosely against the side of a pier, wharf, or bulkhead. They are meant to be struck softly during berthing maneuvers, and have to be replaced periodically. Since they are not driven deep into the mud and are not secured directly to the substructure, fender piles are designed to give considerably, thereby absorbing most of the impact. Ferry terminals employ ferry racks, which create a "funnel" to direct a ferry toward the berth. The racks are composed of several rows of fender piles secured to each other with wood sheathing. The ferry strikes the side of the rack, allows the rack to deflect, and then eases along the rack into the berth. The flexible rack is an important feature because ferries are not very maneuverable and currents near ferry terminals vary considerably with tidal variations and winds.

Strips of reinforced rubber fendering are sometimes used to protect smaller pleasure craft from impact damage. These strips are placed along the edge of the dock to absorb impact. They are also located vertically along support piles or bulkheads to take into account variations in the water height due to tides.

Portable fenders made of plastic, rubber, or similar materials are carried by many vessels for use where no fendering system is available. These fenders are placed over the side of the vessel and absorb impact without allowing the vessel to hit a fixed structure.

For larger vessels, fender piles do not offer adequate protection. At containerports or piers at which large ships dock, large rubber or neoprene bumpers are used. These large blocks are secured by steel bolts and plates to the side of the wharf and are designed to deflect under pressure without allowing the vessel to touch the structure itself. These bumpers come in many sizes and should be matched to the height of the ship's side as well as the size of the ship to be protected. In areas with tides, fender blocks should be located at several heights.

## Siltation and Water Depth

Virtually all shorelines are subject to siltation, the gradual filling of water areas by the transport of organic and inorganic materials by water currents. Siltation rates vary greatly from location to location, but can be as high as 2 feet per year along some rivers. Silt is most likely to be trapped by piers, jetties, breakwaters, or other permanent structures perpendicular to the shore. Vessels that are not moved periodically may actually be locked in place by silt that gathers around the submerged hull.

Proper water depths must be maintained at berthing locations and the channels leading to these areas. The depths must be at least 2 feet greater than the draft of the largest vessel the facility would expect to serve. To maintain adequate depths, maintenance dredging is required at all berthing areas and established channels.

## CONSTRUCTION MATERIALS

The following are descriptions of some of the most frequently used materials in waterfront construction.

## Wood

Wood is strong in both compression and tension and can withstand moderate impact without breaking. It is most often used for pilings, pier substructure, bulkheads, boardwalks, and docks. Wood needs to be preserved in a marine environment, and can be expected to have a moderate service life. Wood can be easily handled in the field with moderately skilled workers and is easy to pile drive into soft soils and silt.

## Concrete

Reinforced concrete has great compressive strength and can be used in applications where tensile strength is somewhat important. Concrete should be installed by skilled workers, and the placement of reinforcing steel is very important, because microscopic surface cracks, which in upland applications would not pose a significant problem, can allow water to enter the material and corrode the steel. Therefore, the steel must not be placed too close to the surface, or spalling of concrete may result. Concrete is used in the construction of piles, pier substructure, pier decks and superstructure, seawalls, breakwaters, docks, and dry-dock walls. With proper maintenance, the service life of concrete can be measured in scores of years.

## Steel

Steel has tremendous compressive strength and is also one of the strongest materials in tension. Steel must be installed by highly skilled labor and can be used for concrete reinforcement, sheet pile retaining walls, bulkheads, pilings, railings, and pier superstructure. Although the cost of the material limits its widespread use, stainless steel can be used in applications in which corrosion resistance is very important, such as vessel fittings and instrumentation. Unprotected structural and decorative steel does not last long in a marine environment and must be protected prior to installation.

## Brass

Brass is a corrosion-resistant metal used for vessel fittings, hardware, latches, lighting, window frames, and other nonstructural applications.

## Aluminum

Aluminum is a corrosion-resistant lightweight metal with a wide range of maritime applications. It is used in constructing vessels, ramps, superstructure, window frames, railings, and conduit. Special marine grades are designed to withstand pitting and corrosion while maintaining structural strength.

## Stone

One of the most durable materials in a waterfront environment, stone is very strong in compression but has little tensile strength. It is used most often in riprap retaining walls, bulkheads, breakwaters, and jetties. It resists wave impact and does not degrade easily. Properly maintained stone structures may last for many decades.

### Rubber and Neoprene

Rubber and neoprene are very resilient and are used primarily as impact-absorbing materials and for corrosion protection of utility lines. Rubber is often reinforced with nylon or steel fibers, and both materials are sometimes reinforced with steel when used as impact-absorbing devices.

### Expanded Foam

Expanded foam is extremely light in weight and is commonly used to increase flotation and as an insulator. It must be protected from ultraviolet light and cannot be placed uncoated in water for extended periods of time.

## DETERIORATION OF MATERIALS

Materials that normally would have a useful design life of many decades can deteriorate much more rapidly in a marine environment. The following paragraphs illustrate several ways in which the water can accelerate the deterioration of construction materials.

### Wood

When kept totally submerged and without exposure to the air, wood can be expected to last for a number of years without significant deterioration. However, when exposed to the air and periodically exposed to rains, submersion in water, or salt spray, wood can rot quickly. The rotting can occur when wood is placed horizontally on a pier deck and water is trapped between the two materials. Water can also attack wood vertically through the end grain. Exposed wood in the splash zone, the area between mean low water and up to 2 feet above mean high water, is regularly exposed to water and deteriorates rapidly.

Rusting connection bolts also have a damaging effect on wood. Wood is also susceptible to impact damage from vessels striking fenders, pilings, and structural wood. Over time, impact damage will require replacement of the wood member.

In areas where the water quality is relatively good and water temperatures are moderate or warm, another danger awaits wood placed in the water. Marine organisms that digest wood can infest unprotected wood and destroy a typical 1-foot-square section in less than two years. There are two basic types of marine organisms that attack wood: almost microscopic, crablike creatures and wood boring worms. These tiny animals feed on the exterior of the wood, relentlessly eating away at the exposed surface until little wood remains. They grow in large colonies and can infest entire harbors in a relatively short time. Some types attack the wood exposed above the waterline, while others eat the wood near the mudline under water. Wood boring worms make tiny entry holes in the surface of the wood and live the rest of their lives boring through the interior of the wood. Eventually, the interior of the wood is weakened to the extent that it is no longer structurally sound. The biggest problem with these worms is the fact that the entry holes are so small that on visual inspection of the outside surface, the wood seems sound.

### Concrete

Over time, concrete pilings can actually disintegrate in salt water. The sulfates in seawater attack the cement binding the concrete together, and slowly erode the cement. The concrete develops an "hourglass" shape as the aggregate crumbles and falls away from the reinforcing material. Eventually, only the rusting steel reinforcing will remain.

Concrete is also prone to spalling, the cracking of the surface of the concrete. This process is started by water attacking the interior steel reinforcing and rusting the steel. Since oxidized steel (rust) takes up several times the volume of the original steel, the rusting process puts tremendous pressure on the concrete as the rust expands. The concrete surface eventually splits and falls away, exposing the reinforcing material to direct contact with the water.

The freeze–thaw cycle can also cause problems with concrete surfaces. Small cracks in the concrete can trap water in winter. The water may freeze, and in doing so, expand and crack the concrete further. This cycle can be repeated many times, the cumulative effect being a deteriorated concrete surface that is likely to spall and crack further.

### Steel

Unprotected steel submerged in water and occasionally exposed to the air will rust relatively quickly in a marine environment. The oxidized steel will rust continuously until little metal is left. This also occurs to steel used in reinforcing concrete. Metal that is continuously submerged will not rust significantly. However, steel placed in salt water can dissolve over time, as metal molecules are stripped from the surface of the steel by metal ions in the water. A significant cross section of steel can be lost in only a few years, weakening the steel substantially.

## Aluminum

Aluminum holds up very well in a marine environment. However, ordinary aluminum will pit severely over time, degrading the surface of the metal and eventually weakening the metal. Marine grades of aluminum, which are alloys of the metal, withstand oxidation much more successfully and can stay in service for decades.

## Ice Effects

Ice can damage structures made of any material. Ice formed around wooden piles can lift the piles straight out of the mud as the ice pack rides up on the surface of the water. Large sections of floating ice can cause impact damage to vessels or piers made of any material and can severely damage floating docks and similar structures.

## DETERIORATION PROTECTION METHODS

The deterioration of materials can be prevented or slowed by utilizing one or more of the following methods.

## Metal Coatings

As a general rule, steel and other metals should not be placed directly into the water. However, if it must be done, several commercial resin coatings offer some help in slowing the deterioration of metals. The biggest problem with coatings is that a small breach of the surface of the coating, through abrasion, impact damage, or incomplete application, can allow water to enter and begin oxidizing the metal. Prior resin coating of reinforcing steel has demonstrated some success in cutting down on the spalling of concrete.

## Cathodic Protection

One means of limiting metal loss due to ions in the water dissolving the metal is to provide a cathodic protection system. Such a system is composed of a cathode, or sacrificial metal ingot, which is suspended by a wire in the water near the metal structure. One end of the wire is attached to the structure, and the other to a source of electricity. A small electric current is sent through the system, and the sacrificial metal is eroded instead of the metal on the structure. For this method to be effective, the cathode must be replaced regularly, and the electrical system must be adequately maintained.

## Drainage Systems

Proper drainage systems are important in eliminating standing areas of water that could rot wood and rust metal fittings on pier decks and bulkheads. Drainage weep holes should be provided at regular intervals along all seawalls and curbing. In addition, wood should never be in direct contact with a pier deck, but should be elevated to allow water to pass beneath.

## Expansion Joints

Proper expansion joints allow the structure to move without cracking concrete decks, dislocating utility chases, or allowing water to collect in the expansion joint itself. There should be a weatherproof seal on the expansion joint, and it should have a cover that is easily removable for inspection and repair.

## Pile Jacketing

The most effective way of protecting pilings is to jacket them. Polyethylene jackets can be wrapped around wood pilings to protect against rot and marine boring organisms. The polyethylene creates a barrier that does not allow air or marine organisms to reach the surface of the piles. Wood or steel piles can be jacketed in reinforced concrete. Formwork is applied to the upper portions of the piles, and concrete is poured within the form. This protects the wood piles from rot while protecting the steel from rust due to salt spray and exposure to the air. The jacket should extend below low water and up to the top of the pile. For added protection, the jacket can be left in place after the concrete is poured. Fiberglass jackets can be used as forms for reinforced concrete pours. These forms are often used on severely damaged piles which must be reinforced without removing the existing damaged pile. The jacket is placed around the pile, and concrete is poured around the form. The form may be left in place to protect the concrete.

## Proper Fendering Systems

Well-designed fenders can protect structures from damage caused by impact from vessels, harbor drift (derelict pieces of piers, bulkheads, and construction debris), and pack ice. Fenders should be situated to keep vessels from hitting key structural elements and should be adequately sized to absorb most of the impact force.

## Weatherproofing Utilities

Utilities should be protected from extreme low temperatures. Water lines should be drained during the winter months if not in use, or if in year-round use, should have heat tracer wire wrapped around the pipes to keep the temperatures above freezing. Insulation would also help keep the water from freezing. Utility chases should also be provided to protect utilities from exposure to salt spray and temperature extremes.

## Ice Protection Systems

Ice breakers can be installed upstream from exposed berthing areas or permanent structures. These breakers are usually made of steel and are designed to allow ice to ride up on a triangular surface until the ice's own weight cracks it into smaller pieces. Water bubbler systems are effective in protecting floating structures and pilings. A system of air hoses and fittings with a series of small openings are placed around each pile and active berth. Air is mechanically pumped through the hoses at a slow rate. The air bubbles prevent the formation of ice, and the berths remain clear of ice. No ice forms on the piles, and there is no upward pressure on the pilings or the attached dock system.

# ADAPTIVE REUSE OF WATERFRONT PROPERTIES

In recent years there has been a significant increase in the number of projects proposed for waterfront sites both in urban areas and on rural shorelines. This resurgence in interest in the waterfront is due to several factors.

As traditional shipping uses along the waterfront declined, owners of break-bulk facilities sought alternative uses for sites that were inappropriate for conversion to containerized shipping operations.

The closing of major landfill sites as they reached maximum capacity created large tracts of land that could often be recycled for public use.

As urban areas expanded, upland sites became increasingly scarce and water-front areas adjacent to prime real estate were viewed as attractive development sites.

Population growth increased the demand for land-intensive uses such as active recreation and entertainment facilities, many of which could be adequately accommodated only at waterfront sites.

Although the cost of developing waterfront sites is usually higher than construction on upland parcels, a number of businesses with unique space needs have begun to utilize underutilized waterfront sites. Others have found that existing waterfront structures can be recycled to meet their needs at a cost comparable to that of renovating space at less attractive inland sites. These new waterfront uses include restaurants, recreation facilities, promenades, offices, performance centers, exhibition spaces, and heliports. Each has unique requirements, as outlined in the following paragraphs.

## Restaurants

Waterfront sites often provide spectacular views of urban scenes or natural landscapes. Restaurants can take advantage of these views to attract patrons, even if the location is distant from existing retail centers and other commercial uses. Eating establishments can utilize relatively small structures that would not be feasible for other waterfront uses, and can be designed to fit within a variety of building types and floor plans. Full utility service is a requirement for the development of a permanent eating establishment.

## Recreation Facilities

In general, recreation facilities require minimal utilities and infrastructure improvements and can be placed at waterfront sites that are physically removed from existing utility lines. Large-scale recreation facilities such as ball fields, tennis courts, stadiums, and pools are often constructed at waterfront sites because of the larger parcels typically found along many shorelines. Such facilities are also appropriate uses for some completed clean landfill sites, which are graded, landscaped, and recycled as recreation areas. Often, parking areas can be incorporated in recreation facilities, which may be precluded at more constricted upland sites.

## Promenades

Pedestrian walks along shorelines have been incorporated in many new projects and provide the most useful and attractive solution to the problem of how a project meets the water's edge. Often, a promenade is constructed as an integral part of stabilization efforts along the shoreline, and may be constructed as the top surface of a bulkhead, retaining structure, or wharf. Food service or other small retail uses are sometimes provided adjacent to the pedestrian path.

## Offices

Waterfront land can be successfully developed as office space if reasonably priced parcels or suitable existing structures are available. Proximity to major highways or existing mass transit routes is a marketing requirement, and adequate parking facilities in suburban areas are also needed. If the site is close to major commercial centers, corporate-headquarter and other high-profile office use can be anticipated. However, when the waterfront site is some distance from an area's commercial center, tenants usually are back-office, incubator businesses or small firms that require moderately priced office space.

## Performance Centers

Public performance centers such as outdoor concert facilities have been constructed on large vacant parcels or pier structures in a number of cities. These facilities either have good mass transit access or large parking facilities, and can be built as temporary structures for seasonal use. Ready access to utilities is required because of the need for electrical power (sound systems, stage lighting), water and sewer service (landscape maintenance, concessions, public rest rooms), and telephones.

## Exhibition Spaces

Exhibition spaces have been installed at both indoor and outdoor facilities in many locations. Outdoor art exhibits of large-scale sculpture and earthworks have been built on vacant land and require only minimal infrastructure improvements or new utilities. Aquariums, museums, and cultural exhibits can be accommodated either in renovated structures or in new construction on land previously underutilized. Such buildings require full access to utilities and usually provide substantial on-site parking. Historic sites are also located along waterfronts and have requirements similar to those of indoor exhibition centers.

## Heliports

Heliports can be constructed on vacant tracts of land in close proximity to commercial or employment centers. Heliports typically require a minimum of 1 to 3 acres of land to provide adequate space for landing pads, offices, parking, and material delivery staging areas. Utility requirements include electrical and telephone service, and where permanent office space is provided, water and sewer connections.

## CONSTRUCTION COSTS FOR WATERFRONT STRUCTURES

It is usually much more expensive to construct substantial buildings or install utility service on the waterfront than to build comparable facilities on upland sites. Construction costs for waterfront structures and infrastructure also vary significantly from one geographic area to another. The following is a summary of typical costs for various waterfront structures. The costs will vary due to live-load requirements, strength of materials, unusual subsurface conditions, local labor rates, and the availability of skilled labor with experience in waterfront construction. Costs will be higher in urban areas and in areas where materials must be shipped long distances or to remote locations. The following ranges indicate the average range of cost for each item installed under normal conditions.

| | |
|---|---|
| Piers and wharves (superstructure) | $100 to $275/sq ft |
| Wood pilings (placed on land) | $600 to $1100/each |
| Wood pilings (placed in water) | $1000 to $2000/each |
| Steel and concrete pilings (placed in water) | $2000 to $4000/each |
| Utility bollards (for marinas) | $500 to $1000/each |
| Wood bulkheads (with lateral tiebacks) | $750 to $1500/lin ft |
| Sheet steel bulkheads | $800 to $1800/lin ft |
| Concrete bulkheads | $1000 to $2000/lin ft |
| Floating docks (wood, metal, or composite) | $30 to $120/sq ft |
| Floating docks (concrete) | $80 to $200/sq ft |

## ADDITIONAL READING

American Association of Port Authorities, *Port Design and Construction*, AAPA, 1964.

American Society of Civil Engineers, *Coastal Structures '79*, ASCE, New York, 1979.

American Society of Civil Engineers, *Ports '77, Ports '80, and Ports '83*, ASCE, New York, 1977, 1980, 1983.

Architecture Research Center, *Port and Harbor Development System*, College of Architecture and Environmental Design, Texas A&M University, College Station, Tex., 1971.

Bradley, E. H., Jr., and Armstrong, J. M., *A Description and Analysis of Coastal Zone and*

*Shorelands Management Programs in the United States*, Ann Arbor, University of Michigan Sea Grant Program, University of Michigan, Ann Arbor, Mich., 1972, Section 72-284, pp. 267–298.

California Department of Boating and Waterways, *Layout and Design Guidelines for Small Craft Berthing Facilities*, Sacramento, 1980.

California Department of Boating and Waterways, *Layout and Design Guidelines for Small Craft Boat Launching Facilities*, Sacramento, 1980.

Cross, Frank L., Jr., ed., *Marine Environmental Engineering*, Technomic Publishing, Lancaster, Pa.

Derucher, K. N., and C. P. Herns, Jr., *Bridge and Pier Protective Systems and Devices*, Marcel Dekker, New York, 1979.

Engineering Foundation, *Coastal Engineering*, Council of Wave Research, The Foundation, New York, N.Y., 1958.

Gunn, C. A., Concentrated Dispersal, Dispersed Concentration: A Pattern for Saving Scarce Coastlines, *Landscape Architecture*, Vol. No. 2, pp. 133–134, 1972.

Ketchum, Bostwick H., ed., *The Water's Edge: Critical Problems of the Coastal Zone*, MIT Press, Cambridge, Mass., 1972.

Marine Technology Society, *Handbook for Offshore Port Planning*, MTS, Washington, D.C., 1974.

May, E. B., *Environmental Effects of Hydraulic Dredging in Estuaries*, Alabama Marine Resources Bulletin 9, Department of Conservation and Natural Resources, Montgomery, Ala., 1973.

McAleer, J. B., C. F. Wicker, and J. R. Johnston, *Design of Channels for Navigation: Evaluation of Present State of Knowledge of Factors Affecting Tidal Hydraulics and Related Phenomena*, Report 3, U.S. Army Engineer Committee on Tidal Hydraulics, Vicksburg, Miss., 1965.

National Fire Protection Association, *Control of Hazards on Vessels to Be Repaired*, NFPA 306-1975, NFPA, Batterymarch Park, Quincy, Mass., 1975.

National Fire Protection Association, *Fire Protection Standards for Marinas and Boatyards*, NFPA 303-1975, NFPA, Batterymarch Park, Quincy, Mass., 1975.

National Fire Protection Association, *Operation of Marine Terminals*, NFPA 307-1973/1975, NFPA, Batterymarch Park, Quincy, Mass., 1975.

National Fire Protection Association, *Standard for Construction and Protection of Piers and Wharves*, NFPA 87-1975, NFPA, Batterymarch Park, Quincy, Mass., 1975.

San Francisco Bay Conservation and Development Commission, *San Francisco Plan*, The Commission, San Francisco, January 1969.

Schenck, Hilbert, Jr., *Introduction to Ocean Engineering*, McGraw-Hill, New York, 1975.

Schenker, and Brockel, *Ports Planning and Development as Related to Problems of U.S. Ports and Coastal Environment*, Cornell Maritime Press, Centreville, Md., 1974.

Simmons, H. B., J. Harrison, and C. J. Huval, *Predicting Construction Effects by Tidal Modeling*, Miscellaneous Paper H-71-6, U.S. Engineer Waterways Experiment Station, Vicksburg, Miss., 1971.

U.S. Department of Navy, Bureau of Yards and Docks, *Design Manual for Harbor and Coastal Facilities*, U.S. Government Printing Office, Washington, D.C., 1948.

Water Resources Council, Proposed Principles and Standards for Planning Water and Related Land Resources, *Federal Register*, Vol. 36, No. 245, Part 2, pp. 24,144–24,194, December 21, 1971.

Wiegel, Robert L., *Oceanographical Engineering*, Prentice-Hall, Englewood Cliffs, N.J., 1964.

## QUESTIONS

1. Describe the four ways in which water bodies meet land.

2. Describe how winds affect both floating and immovable objects along a waterfront.

3. Identify the two most significant problems encountered when developing on landfill sites.

4. Discuss the differences between piers/wharves and platform structures.

5. Describe the various building techniques and materials used in the construction of bulkheads and retaining walls.

6. Describe the three types of bearing piles, including their applications and methods of installation.

7. Describe the three key issues affecting the stability of floating structures and two methods of stabilizing floating structures.

8. Describe the four types of fendering systems used to protect floating vessels from direct contact with stationary structures.

9. Describe two methods of protecting steel from deterioration when the metal is placed in water.

10. Describe two types of ice protection systems for stationary waterfront structures.

## GLOSSARY

**APRON**   Clear area around perimeter of a dock for parking, storage, work areas, and access.

**BERTHING BASIN**   Area of a harbor set aside for berthing vessels at docks and/or open anchorages.

**BULKHEAD** Structure designed to retain earth or fill and consisting of a vertical wall sometimes supported by anchors and tierods upland of the bulkhead.

**BULKHEAD LINE** Establishes limits outside which continuous solid-fill construction is not permitted.

**CAISSON** Watertight box used to surround the construction site, work crews, and equipment used to build foundations for bridges or other structures below water.

**CAMEL** Float placed between a vessel and a dock (or between vessels) designed to distribute wind, current, and impact forces. Camels also keep vessels from coming into contact with pier structures or other vessels.

**CHOCK** Horizontal component of a fendering system used to brace the vertical piles or fenders. Also refers to a mooring fitting used to guide lines.

**COFFERDAM** Temporary wall serving to exclude water from any site normally under water so as to allow the laying of foundations or other construction work.

**CONTROLLING DEPTH** Minimum depth of the navigable portions of a waterway governing the maximum draft of vessels that can enter.

**CURRENT** Flow of water.

**DEADMAN** Concrete, steel plate, or other heavy anchorage for a land or water tie.

**DEPTH OF VESSEL** Distance measured from the main deck of a vessel to the bottom of the hull.

**DOCK** Pier or wharf used for berthing vessels and for the transfer of cargo and/or passengers.

**DOLPHIN** Structure, usually consisting of a cluster of piles, placed near piers and wharves or similar structures to guide vessels into their moorings, or to fend off vessels that may strike structures, shoals, or the shore.

**DRAFT** Depth of a vessel's hull below the waterline.

**DREDGE LINE** Line establishing the limit of dredging.

**EBB TIDE** Period of tide between high water and the coming low water; a falling tide.

**FENDER** Device or system placed against the edge of a dock to take the impact caused by berthing a vessel.

**FLOOD TIDE** Period of tide between low water and the coming high tide; a rising tide.

**FREEBOARD**  Distance between the weather deck of a floating vessel and the waterline.

**HARBOR**  Usually refers to a sheltered arm of the sea, easily accessible to marine routes, in which ships may seek refuge, transfer cargo, and/or undergo repairs.

**HARBOR LINE**  Controls the location of shore structures in or adjacent to navigable waters.

**LEE**  Shelter, or the part or side sheltered or turned away from wind or waves.

**MEAN HIGH WATER (MHW)**  Average height of high water over a 19-year period, or an equivalent calculation of that level.

**MEAN LOW WATER (MLW)**  Average height of low water over a 19-year period.

**MEAN SEA LEVEL**  Average height of the surface of the sea for all stages of tide over a 19-year period, usually determined from hourly height readings.

**MOLE**  Massive, land-connected, solid fill structure of earth, masonry, or large stone (also called a "groin"). It may serve as a breakwater or pier.

**PIER**  Dock that is built from the shore out into a harbor and used for berthing and mooring vessels.

**PIERHEAD LINE**  Establishes the outboard limit for open pier construction.

**QUAYWALL**  Heavy gravity or platform structure fronting on navigable water, behind which earth fill is placed to a level grade along its entire length.

**RELIEVING PLATFORM**  Platform supported by piles, employed in wharf construction to relieve lateral pressure from surcharge.

**REVETMENT**  Facing of stone, concrete, or other form of armor built to protect an embankment or shore structure against erosion by wave action or currents.

**RIPRAP**  Facing layer or protective mound of stones randomly placed to prevent erosion of an embankment or structure; also the stone used in its construction.

**SEAWALL**  Structure built along and parallel to a shoreline for the purpose of protecting and stabilizing the shore against erosion resulting from wave action.

**SLIP**  Space between two piers for berthing a vessel.

**SPLASH ZONE**  Range between mean low water and the upper limit to which the structure could be expected to be wetted by average wave disturbance at the site.

**STRINGER**  Longitudinal member in a structural framework.

**SWELL** Wind-generated waves with a longer period and flatter crests than waves retained within their fetch.

**TIDAL RANGE** Difference in height between consecutive high and low waters.

**WALE** Horizontal component of a fendering system generally placed between the vertical fenders and the pier structure to distribute the forces from a vessel horizontally.

**WATERLINE** Juncture of land and sea which may migrate due to fluctuation in the water level. Also refers to the line to which water reaches on the side of a vessel or structure.

**WAVE** Ridge, deformation, or undulation of the surface of a liquid.

**WAVE HEIGHT** Vertical distance between a crest and the preceding trough.

**WHARF** Dock, oriented approximately parallel to the shore, used for berthing or mooring vessels.

Newark Airport

# RAIL/TRANSIT AND AVIATION

**Sarelle T. Weisberg**
NYC Dept of General Services

## INTRODUCTION

Rail and aviation infrastructures each have a distinct and critical role to play in a region's and a nation's transportation system network. Both act as magnets for economic health and economic growth for urban areas. Rail systems have an old and well-understood technology, are land-based, are visible, and in the case of subways and local rail lines, can be accessed by walking or using short bus links. Subways play a special role in the daily business life of the city and contribute to vitality and mobility within a city's central business district (CBD). The routes of a rail system, a fixed guideway on a dedicated right-of-way, create a new geometry for a developing city or region, and play a primary role in determining the urban form that development will take. Land use and land values have a dynamic interaction with growth patterns and increased real estate values, whether a rail line is interwoven in the fabric of the inner city or is stretching out into the suburbs beyond the city core.

In contrast, aviation is a younger system, relying on more advanced technologies for equipment and control. As an air-based network, its travel lanes are invisible, but it is tied to a land-based system of airports and to the local ground transportation linkages. Unlike rail systems and subways, aviation does not create an urban geometry and its airport system is often in conflict with its neighbors and neighboring land uses for its potential for decreasing land values in the immediate vicinity of busy airport hubs. Aviation, as an international network, differs from the more local or national rail systems, in that it must coordinate and cooperate across the skies of many political jurisdictions. Not all of the skies are "friendly," yet today's international trade demands an infrastructure that is reliable although operating in hostile world environments.

Ground-based rail systems can face jurisdictional conflicts as they grow to link areas beyond local townships and across county or state lines. Government intervention, including legislative creation of super agencies or dedicated authorities, is needed to ensure the desired equity and continuity of services for the benefit of the larger society, a traditional policy for infrastructure development and maintenance.

Most significantly for network development and infrastructure renewal, the older rail systems today cannot be supported by fare box receipts and therefore, in the U.S. example, require a high level of local and national subsidy (Urban Institute, 1987). Ridership has been shrinking on the most neglected of our older lines. Operating expenses, particularly for minimally automated lines, have climbed much faster than any fare increases. To retain ridership and to serve lower- and medium-income riders who have no other transportation choices, it is politically necessary to maintain low fares, even in the face of increasingly less profitable operations.

Coupled with the need to keep fares low is the extraordinary degree of neglect in maintaining the old systems, which will take a level of resources to restore that may not be available in a period of increased competition for national funding. For years, municipalities used their maintenance budgets to fill the gap in their operating budgets, relying on deteriorating systems to perform years beyond their original expected life.

Unlike the rail mode, aviation's major users in the United States do pay for the system's development and maintenance. Airline ticket taxes are the major source of revenues to airports and also support the Federal Aviation Administration (FAA) and its air traffic control responsibility. Aviation is considered a growth industry, in contrast to rail services, which have either experienced diminishing markets in older cities or lower densities that do not yet support their newer automated systems in growing metropolitan areas.

We may define aviation infrastructure as relatively independent in its ability to finance its needs. It is, however, a highly dependent infrastructure within the family of transportation modes, relying as it must on local highways and rail linkages to complete the trip, which is really a multimodal experience. It can take more than an hour (depending on peak-hour delays) to access the system at either end, substantially affecting time and cost comparisons for flying.

In socioeconomic terms, aviation remains the mode of a more affluent society. Its passengers include a high component of business travelers who can afford to pay for the services they want, unlike subway riders. The subways of the world must function across the full economic spectrum and provide even the lowest-income user with an affordable daily commute.

Technical challenges in transportation infrastructure are significant. Older rail systems have inflexible routes and are difficult to expand. Tunnels and bridges present real constraints in updating and upgrading of older lines. New technologies are usually not easily superimposed upon existing track and communications networks. It is also complex to shut down large segments of any system to build the needed improvements; disruption and dislocation are very costly. The phasing of improvements is both technically complex and fiscally hazardous. Phasing must match the need with available resources in an annual budget cycle, over at least a decade, since transportation projects require a very long time frame from planning to realization.

For aviation, research and development must focus both on technical obsolescence and requisite compatibilities between aircraft and airports or between first-generation versus advanced electronic control systems. Aviation is more closely tied into the research laboratory than are rail systems; the aviation technology is constantly improving, in some measure in response to defense requirements.

Both rail and aviation modes have an active goods movement component, with air cargo today increasing to meet changing profiles in business and manufacturing worldwide. Rail cargo faces intense competition from highway trucking, which can offer more flexible routes and some point-to-point deliveries. The discussion that follows will focus primarily on the infrastructures as passenger services, in their role in meeting the social and economic goals of the urban community.

# RAIL AND TRANSIT INFRASTRUCTURE

## Role of Rail and Transit Infrastructure

Rail infrastructure dates from the 1860s in Europe, Asia, and the United States and has a long history as an old and venerated technology based on the steam-driven locomotives of Liverpool and Manchester of the 1830s, electrification for suburban trains by 1900, and the arrival of diesels by 1925 (Georgano, 1972). The major physical attribute of rail systems is the visibility of this primarily land-based network, which as it reaches beyond the city core, creates an urban geometry for the city and its surrounding suburbs. In the early years, the middle and late nineteenth century, rail systems united the older European capitals, tied together great continents spanning vast stretches of undeveloped land, and were of extraordinary political and nationalistic significance. Built by the great entrepreneurs of the new industrial age, railroads symbolized the aspirations of capitalism and world leadership for the future, much as the airlines of today embody a nationalistic aura beyond the statistics of the passengers they carry.

As urban densities increased, and more people were drawn to centers of commerce and industry, suburban developments became increasingly important as residential relievers of the urban densities. The suburbs of Chicago, Philadelphia, Boston, and New York were eased by the developments in Oak Park, Main Line Philadelphia towns, Roxbury, Brookline, and Brooklyn.

In older downtowns at the turn of the century, the elevated loop or linear line was woven into the existing district, and in Chicago toady we ride the original core segments of the "Loop," now linked to outlying communities. The early "el's" and their extensions created the opportunity for real estate development, affecting land use and land values, similar to the role played by post–World War II highway programs in developing the housing subdivision tracts of the Levittowns USA.

In today's urbanized suburbs, it is clear that the economic health of a developing region depends on the transportation system networks available to maintain the mobility of the work force, to attract new residents and commercial enterprises, and to increase the pace of land development. New transportation links have created vital corridors and hubs for growth at modal interchange points, park-and-ride sites, and the new stations. The systems expanding beyond the city, built in the 1970s and 1980s, contribute to the decentralization of the older cities, a trend now predicted to become one of the major urban characteristics in the United States by the year 2000.

Where the local political and governmental authority is limited by geographic boundaries, the difficulty in establishing an appropriate development and operating authority that can transcend the jurisdictional problems for ground-based systems is one of the prime issues in transportation systems planning. The degree of cooperation and coordination required in developing networks that must cross and recross many county, town, and state borders to maximize the network potential for its users and operators remains one of the major planning frontiers.

The great surge of suburban sprawl in the 1940s and 1950s in the United States was due to the extensive highway development program. While existing suburban rail lines provided an option for the commuter to park at the station and complete the trip downtown by a public transit mode, this was only true in the highest-density areas and for the trip into center city. Interconnections among suburban centers was wholly dependent on the private auto, as was the commute to the city in all but a few major metropolitan regions in the United States.

## Service Classifications

Some confusion may exist in the definition of public transit, mass transit, and suburban rail transportation. Public transit includes a family of modes: trolley, tram, bus, cable car, funicular, or a combination of these scheduled and frequent services. Public modes can include taxis or limousines, which require a public franchise or operating authority but do not provide a high occupancy or group ride. Mass transit refers to the subway and metro systems with multicar operations, carrying 100 persons per car, and possibly 10 cars per train. Mass transit serves a very concentrated and dense urban area.

Several general classifications used to identify urban transit infrastructures, either public transit or mass transit systems, are shown in Table 9-1. Each classification has a distinct historical place in the development of the city and suburb; some services have fallen in and out of favor and public acceptance only to reappear to meet some specialized new mobility need. The least flexible service is the heavy rail subway, which may run partially above grade and underground, providing local and express service in a semiautomated system using electric power. Light rail and some medium rail systems can be adapted to difficult terrain and threaded within an existing urban context, but if special and dedicated rights must be secured, once in place, there is less flexibility.

**Table 9-1   GENERAL CLASSIFICATION OF URBAN TRANSIT INFRASTRUCTURE**

Geographic location
  *Service area:* inner city, suburban, intrasuburban, intercity
  *Route type:* linear, radial, loop, combined
Physical location
  *Surface:* segrated, grade-separated, dedicated right-of-way (ROW)
  *Above grade:* usually on dedicated structure, on dedicated ROW on bridges, point-
    to-point inclines, cable cars
  *Below grade:* tunnels that are shared, segregated, or dedicated
Service delivery characteristics
  *Local:* all stations on the line
  *Express:* skipping local stations, stopping at interchange points
  *Limited:*   special services in peaks, with designated stops only
  *Combined:* all of the above
Power systems
  *Type:* steam, electric, diesel, air
  *Location:* rail, monorail, cable, track bed
Control systems
  *Manual:* labor intensive, old systems
  *Semiautomated:* older systems, modifications
  *Fully automated*
Vehicle class
  *Heavy rail:* subways
  *Medium rail:* trolleys, trams, passenger distribution systems (PDSs)
  *Light rail:* Commuter lines, PDSs, street cars

## Modal Choices

Research on modal choices in this country has reinforced the finding that except in periods of energy crises and high automotive fuel costs, public transportation alternatives do not compete favorably with the auto, despite known energy savings (Ruppethal, 1981):

|  |  |  |
|---|---|---|
| Average auto | 1.3-person capacity | 6897 Btu per passenger-mile |
| Subway | 100 capacity/car | 205 Btu per passenger-mile |
| Subway | 200 capacity/car | 103 Btu per passenger-mile |

It can be argued that the energy costs to build a new system should be factored into this energy cost, but experts suggest that between 5 and 30% of the investment in rail transit can be recouped by the energy savings over time. In a marginal cost analysis, it can be shown that existing public transit and mass transit systems can accommodate additional ridership relatively inexpensively, even in peaks, by adding services to an in-place network, thus validating the energy comparisons.

The maxim says, "As automobile ownership increases, transit demand declines." Coupled with the highway expansion program, this was substantially true, but our present highway congestion plus the intensified growth beyond the cities has produced peaking densities in newer cities and suburban communities that seem to require a transit solution, even where present densities may not yet have forecast that need to planners (Pushkarev and Zupan, 1980).

As we examine the history of our older rail and subway networks, a prevailing "truth" should be acknowledged: the attractiveness of public modes has only a minimum effect on auto use, whereas the restraining of auto use, by policy and enforcement, does have a very perceptible impact on transit use.

We are witnessing the rejection of old and poorly maintained rail infrastructures in favor of expensive reliable express bus service in the rush hours of major cities. Improved attractiveness of older systems is critical for three major trends: (1) in "headquarter" cities there is a centralizing of white-collar and professional employment originating in outlying suburbs and dependent on suburban rail plus inner-city subway links; (2) decentralization is beginning for manufacturing and distribution jobs in and around successful suburban centers which are still auto dependent, but a large inner-city labor supply is in need of getting there; and (3) a new concentration of "back office" development by major corporations tapping into the white-collar labor pool has few public transit commuting options across the suburban landscape, but the current densities do not support fixed-guideway systems and ridership is not established along one specific corridor. Alternatives to fixed-guideway rail systems fall into the transportation systems management (TSM) category, to intensify the use of the public systems already in place, such as highway initiatives of contraflow lanes and high-occupancy vehicle lanes for vans and carpools, or staggered workhours within the CBD to ease subway peaks. The point at which rail modes become the superior modal choice for the commuter as well as, say, the airport-bound passenger, depends on the availability of separated rights-of-way for competing highway modes versus the service reliabilities of fixed-guideway rail.

## Performance Criteria

Selecting a service mode, passengers rate the auto their first choice except in dense urban areas where time delay, parking expense, and the availability of a genuine alternative urban transportation infrastructure exists. Patrons are less price conscious than quality sensitive. Evaluations and comparisons by both transportation professionals and patrons are made on the basis of a wide range of perceptions and experience that place convenience and reliability factors above cost. Performance criteria include:

| For the Patron | For the Professional |
| --- | --- |
| Schedule | Frequency of service |
| Travel time | Operating speed |
| Reliability | Reliability |
| Safety and security | Safety |
| Peak-hour seating availability | Line capacity |
| Value for fare | Output per unit of fare productivity |
| Availability off-peak | Utilization |
| Convenience and location of stations | On-line facilities |

In the 1980s in the United States, although not all experts agree, only about 6% of commutes to work were via transit, down from the 9% associated with the 1970s. This statistic does not characterize major metropolitan regions, such as New York/New Jersey, where in 1981 over 30% of all U.S. subway trips were made. This region is considered the most transit-dependent area in the country (Port Authority of New York and New Jersey, 1985a). On a typical day, over 50% of work trips, 60% of all trips, and 80% of CBD work trips are by public transit (including bus transit).

In this family of personal transportation choices, each mode, from the private auto to rail rapid transit, has its optimal domain of application. Closely related modes should have an overlap of ridership, but the distantly related modes should not be competitive, but should operate in complementary roles to form a total and coordinated transportation infrastructure network.

This need for a balanced system of several complementary modes was recognized during the post–World War II period in Europe, when cities were rebuilding their bomb-torn centers and highways and autos were not the selected priority. Two significant decisions of that time contribute to today's high level of coordination of on-time integrated fare-transfer systems among local trams, buses, metros, and regional rail (Cooper-Hewitt Museum, 1977). First, the understanding that the total network must be organized as a single entity operating with a high degree of reliability, noncompeting, with costs maintained by sustained levels of government subsidy as a necessary base for economic recovery and health, was accepted. Second, a sound economic basis was established that included up-front recognition that maintenance costs must be on a pay-as-you-go basis and as in private-sector for-profit operations, the investment in public works must not be allowed to fall into a pattern of deferred maintenance, with resulting loss in system performance and reliability.

### Densities and Demand

In Roman times, and throughout the nineteenth century, densities in cities remained at 20,000 to 40,000 people per square mile. Growth in cities by the beginning of the twentieth century produced densities of 60,000 to 95,000 people per square mile. In 1910, the average Manhattan density reached 130,000 people per square mile, with levels of 280,000 to 520,000 in walk-up tenements filled with new immigrants. In New York, Prague, Paris, and other dense urban centers, the advent of the electric street cars and rapid transit helped to dissipate this extraordinary condition (Vulchic, 1981). In England in the early and mid-nineteenth century, the newly invented steam lines took London workers out into the gardens and countryside for relief, but lack of inner-city transit resulted in maintaining similar extraordinary crowding and very poor living conditions within the old city.

Steam railways in the United States provided local service via intercity rail links used by relatively few affluent businesspeople commuting to stately homes in the suburbs. Some of this class stratification still exists in the metropolitan suburbs, where, for example in the New York/New Jersey metropolitan region, the more affluent settled in Long Island, Westchester, Rockland, and New Jersey's Morris County and commute on newer, better-maintained electrified rail systems, while the older, "poorer" subways were neglected, operating within the city where lower-income residents are dependent on them for their daily work trips.

Until the 1950s, planners thought that a population of 1 million represented a threshold for developing rail rapid transit systems. While a city population of less than 500,000 would be considered the exception to support a metro, cities of 750,000 or 900,000, for example Stockholm and Lisbon, have built successful services (Cooper-Hewitt Museum, 1977). Light rail systems are usually planned for urban areas of 200,000 to 300,000 population. For cities in the range of 500,000 to 2,000,000, light rail and subway systems are in place, such as in San Francisco and Toronto. Factors other than population levels can make a success of a line, particularly if there are auto disincentives such as high tolls and parking fees awaiting those who drive into the CBD, versus a well-run reliable alternative rail system in their suburban corridor.

## Peaking Problems

Peaking problems in the urban transportation systems are more serious than those occurring in the public utilities, since if you raise peak-hour fares, you will discourage use of transit and lower the very density on which the system depends. The public mode then competes with the private auto mode, to the ultimate detriment of the public mode. Unlike the utilities, where there is no alternative available (except for the exceptional cogenerator, perhaps), the public transportation rider frequently can choose an alternative.

Operators of the transit systems focus on maintaining service comfort during peaks by spreading peaks into perhaps a 3-hour period, instead of squeezing them into a $\frac{1}{2}$- or 1-hour time slot. Strategies include using extra trains, tighter schedules, adding special express and limited stop services, and working with the business community toward a pattern of flexible or staggered work hours to help to flatten peak-hour demand to the system. Off-peak slumps are made more attractive for nonbusiness riders by lowering fares for special user groups: children, seniors, and so on.

For public transit professionals in the United States today, the goal is to increase ridership by attracting new and higher market shares, preserving optimal cost competitiveness with private bus operations and private autos. At any given cost per passenger, more service per unit area can be provided at higher density, because demand is higher. Conversely, at any given service level, cost per passenger is lowered with the rising density, which translates into lower fares, which in turn helps to attract additional passengers to the system and thereby improves service per unit of area served.

## System Components

Components of the rail and transit infrastructure system include rights-of-way (R/Ws), stations, and maintenance facilities (Vulchic, 1981). The *right-of-way* is the strip of land legally owned by the transportation facility, on which the system runs. Different types relate to different track positions: elevated or underground. The geometric elements of the right-of-way have three aspects: (1) horizontal alignments, in which the limiting geometries are the minimum curve radii; (2) vertical alignments or vertical track profiles, which are limited by the maximum acceptable gradient; and (3) clearance or free cross-sectional area for vehicle passage, particularly important for tunnels. The *roadbed* is the earth base on which the track superstructure rests. The bed can be a simple earth one or a major structure in a cut-and-fill system. Crushed stone ballast is most common for at-grade or embankment alignments.

*Way structures* refer to all structures on the right-of-way: bridges, viaducts, aerial structures, and tunnels. *Superstructures* are all fixed physical components that directly support and guide rail or (fixed-guideway) vehicles. They must exist along the entire way length. The *guideway* for rail systems is the track itself. The guideway network must include complex switches. To achieve the necessary operational flexibility and high level of service, transit guideways should always have an adequate number of sidings, turn-back tracks, and crossovers. *Wayside engineering* includes signals, communication cables, and storage facilities located as needed along the line. *Electrification equipment* refers to all elements needed to bring power from the power plant to the vehicles on the tracks, including substations, third-rail, grounding, and return-current conduits.

*Yards and shops* are generally located in outlying areas both for economy in finding large sites and for environmental reasons. These facilities include shops for regular maintenance, inspection, and minor repairs of cars and for cleaning and washing of cars. Tracks for overnight storage of cars, tracks for cars waiting for maintenance, and tracks for maneuvering and preparing trains are also necessary in these yards. General overhaul and major repairs are usually done at one major shop for the entire system. Location should be determined to minimize empty runs.

*Control centers* are major communications centers, "war rooms" for operational and monitoring functions. As more and new systems are automated, the centers control signal and switch setting, and have on-line computer control of operations, with the ability to compute schedule delays and to recover the delays within the system by regulating on-line speeds. Automation's major function is servicing and scheduling of the system; a minor function is to provide accurate and timely passenger information.

*Stations and terminals* represent the contact points for the system. Connections are made with other modes or other transit services. At these modes strong interaction occurs with environments, often making these elements a major investment in the system. It is here that the system strongly affects passenger convenience, comfort, and safety, all dependent on service reliability, operating speed, and line capacity. If an interchange point is developed for different rail lines, these lines can be brought into the terminal on parallel platforms, or can be stacked one above the other, simplest if they are part of the same line. In a more complex station configuration, lines may be crossing at different angular geometries, or the planning criteria may call for passengers to be guided by distinct corridors to each transfer. In these more challenging designs for station interfaces, often where new systems are grafted onto older services, fare collection barriers also interfere.

*Station platforms* can be developed either with a central platform serving trains in both directions or on a lateral plan where each direction has its own separate platform. The advantages of the central plan are: the economy of needing only one set of facilities to serve all passengers; supervision manageable by a single person or TV monitor; passengers can reverse if they have made a mistake in direction without leaving the platform; and the width of the station itself can be narrower since there will be peak activity in one direction paired with off-peak activity in the other. Lateral or decentralized platforms must be wider, designed to take peaks in each direction, in addition to the disadvantage of duplication of fare purchase and other services. Also, security and safety is diffused when split platforms are used.

The most typical station design has three levels: the street or building level, with entrances and exits; the middle or mezzanine level, where fare collection occurs, usually partially a "free zone" and partially a paid zone; and the third level, platform and tracks. Developers today are contributing more to the infrastructures within their projects, so we see a variety of design solutions integrating station design into large commercial and business projects at station points in the system (Urban Land Institute, 1987).

## Funding and Costs

In the assessment of resources needed for the renewal of the older U.S. infrastructure systems, the transportation sector of highways and transit systems are expected to experience a minimum of 50% shortfall between anticipated available resources and identified needs. The combined factors of the age of the transit systems, the years of neglect and deferred maintenance, the technological obsolescence of communications and controls, and the overriding reality that operating revenues rarely cover 50% of operating costs serve to place this infrastructure in jeopardy of reaching its potential as the solution to urban and suburban congestion.

In the United States the estimated annual need for infrastructures through the year 1990 is at a level of $53.4 billion. Transit is identified as needing 10% or $5.5 billion nationally, over three times the estimate of need in the aviation sector, and far below the 52% or $27.2 billion needed for highway improvements (Congressional Budget Office, 1982).

Within the projected allocation for transit funding, the largest commitment, 68%, is to be spent for improvement to physical facilities; vehicle replacement will use 23% of the funds, and improvements to maintenance facilities will use the remaining 9%. The age of the vehicles and maintenance facilities is extraordinary: over 25 years for the vehicles and 50 years on average for the maintenance plants.

Cost breakdowns for public transit can be identified under five categories:

1. Transportation, which includes the wages and salaries of all personnel and the materials needed by operating personnel
2. Permanent-way maintenance, which includes the cost of personnel and materials needed to maintain the tracks, power supply, and signals
3. Vehicle maintenance, which encompasses expenditures for personnel and materials necessary for maintenance, repair, testing, and cleaning of cars to meet legal standards
4. Power for electrical cost
5. General and administrative costs covering all indirect operating costs, management, legal, accounting, insurance, benefits, maintenance of buildings and grounds, and other miscellaneous costs

The American Public Transit Association (APTA) uses a somewhat different breakdown, from an operating perspective, so that the transportation item represents almost 50% of their projected needs, with vehicle maintenance taking another 30% of available resources (American Public Transit Association, 1984).

The cost of construction in the transit infrastructure, particularly for new systems or new extensions to existing systems, is generally related to the extent of below-grade and cut-and-cover work in the project. In 1982–1983, Boston's Red Line extension of 3.5 miles was estimated at a cost of $620 million, but this $177 million per mile included twin underground tunnels, four new stations, a 2000-car parking garage, public landscaping, and urban design amenities. In the 1970s, Washington, D.C.'s and Baltimore's new metros were estimated at $26 and $23 million per mile for the tracks and stations.

Operating costs are extremely varied, since there are so many different reporting systems, based on costs per kilometer or costs per car-hour. In all cases, however, expenses of public transit systems increase more rapidly than inflation and revenues grow at a much lower rate. Labor costs to older nonautomated systems are one of the major expense items. Thus the gap between operating revenues and expenses continues to widen, requiring an ever-greater percent of subsidy to maintain a high level of service at a competitive (low) fare.

Vehicle costs are analyzed in terms of life-cycle and maintenance-cycle expectations. A heavy railcar has an estimated life of 30 years; after 20 years a major rehabilitation may add 10 or 15 more years to the car's life. Once a car reaches 30 years, an evaluation will determine if a second rehab cycle can add an additional 10 years of useful life.

The initial cost of a new heavy railcar is now over $1.1 million. Major rehabilitation costs at least $400,000; minor rehab is in the range of $50,000 to $100,000. In recent years car costs have accelerated in a seller's market since there are very few heavy rail manufacturers available. Over the last 25 years, 50- or 51-foot railcars have accelerated in cost tenfold: from $115,000 to over $1,100,000.

In the United States, to provide the needed subsidies to transit, frequently in the 50 to 60% range, a variety of local strategies have to be in place. Operating costs are not eligible for federal Urban Mass Transportation Administration (UMTA) funds. A small allowable trade-off in capital dollars will disappear in 1989.

In the past, UMTA has provided federal money to the 75% level, for capital and demonstration projects, but this is threatened annually by budget realities, to reduce deficit spending. The trend is toward local priorities being dealt with through local enterprise (Urban Institute, 1987). In some states, bonding appears annually on the ballot, but this voter dependency makes long-range planning uncertain until the available resource level is known.

A bistate proposal in the New York/New Jersey metropolitan region would establish an infrastructure bank as a revolving funding source into which developers, in recognition of their dependency upon local infrastructure, are obligated to contribute. The fund would be available for both new services and for the maintenance of existing services that will serve either the project or the region in which the project is to be located. It is, in concept, related to the impact fees imposed for land development that either take care of the requirements of the new project itself or can be part of the developer's contribution to the needs of the larger community.

## Privatization

In the early years of America's expansive growth, private entrepreneurs developed our rail lines, the industrial wealth of the east coast and the middle west providing the capital to forge the rail link between east and west. Transportation was a profitable business and a growth enterprise by the early twentieth century. As competition increased and profits declined, however, government was forced to step in to ensure a transportation network that was clearly an essential service to the nation even though it could no longer pay its own way. Transportation systems are basic, the underpinning to reaching a society's economic and social goals: bolstering the local economy, enhancing regional mobility, and contributing to the quality of life by providing access to opportunities for jobs, education, and culture, while decreasing environmental pollution.

In 1984–1985 the UMTA embarked on private-sector initiatives to encourage joint development of new transit service lines. Stations, people movers, and pedestrian amenities were privately financed up front to maximize the land-use opportunities and would later be ceded to the local transportation authority (Urban Institute, 1987). In the Dulles Airport Rapid Transit example, the federal government supplied the land needed for the new alignments and supervised the construction, but private initiative and private funds made it happen. Rice University studies have shown that it can be 35% less expensive for the private sector to construct public works than when the public sector builds. The long-awaited cross-channel tunnel between England and France will have major private-sector financing initiatives.

Following are the advantages and disadvantages of private development or operation of a transportation infrastructure, in contrast to when the public sector is solely responsible.

| Advantages of Privatization | Disadvantages of Privatization |
| --- | --- |
| More flexible response to changing needs and services | Difficult politically to eliminate low-use/unprofitable services |
| Greater leeway in settling labor issues and contracts | Political opposition by unions and others to loss of jobs |
| Can go in and out of business | Must retain a range of services |

| | |
|---|---|
| May risk new technologies in small specialized systems | Scale of operations, minimum technical advancement |
| Skims off most-profitable routes and markets | Left with least profitable services that government must provide |
| Specialization, segmentation for profit | Loss of control of overall network |
| Can attract private equity capital | Must ask electorate for bonding or raise taxes |

Hybrid solutions can be found today, where public ownership is retained but operating contracts are given to the private sector, particularly for systems requiring minor capital investment in fixed guideways or facilities, such as public transit bus routes. Private operators can come into and leave the system with no front-end long-term commitments of the type that governments must make to maintain and build the necessary service infrastructures.

## Urban Design Issues and Impacts

Transportation systems are the most powerful means of affecting the design of the city. Of all the infrastructures, transportation has the most direct interaction with land use, and consequently with land values. It is often identified as the critical underpinning of economic development.

The skeletons of modern cities were tightly attached to street car lines and railroad stations. New suburban growth is occurring along the outlying spurs of metro extensions and lining once-industrial river edges of older established communities along the commuter rail lines.

In the older dense city core, it is very difficult to insert a new system or to expand an existing service since it is both costly to construct and costly in the disruption of the life of the city parallel to or above the new line. Even a new underwater tunnel, in an unobstructed space below the river, creates disruption at the land connection at either end and can be expected to produce community delaying tactics and be slow to receive all necessary environmental and safety approvals.

In newer and developing cities, at issue is whether a new transportation system will create a centralization and density resulting in undesirable over-crowding, although without the system the city cannot grow. Conversely will a new highway or outlying transportation connection create a shift away from the center and thereby pull the activity away from the core and diminish density and the use of the transportation improvement by decentralization?

Small or new cities often have one dense corridor that will support a linear public transportation system, either bus or light rail. In large older cities, public transportation has been emphasized in their early planning and implemented as they grew and experienced waves of economic development.

The inflexible route of urban rail systems requires a very long range policy commitment to land developments and some assurance of growth. These routes can, and do, determine a city's form and geometry, in consort with the natural topography and land formations. The triumphant boulevards and ceremonial intersections of Paris mark a pre-transit-planning era; the outlying patterns of linear development and modes of activity determined modern Montreal's urban form and growth.

Where activity increases, at stations and intermodal exchange points, commercial uses are attracted and land values increase. If there is an elevated section in the system which can be more easily threaded through developed areas or existing cores, land is freed below the system for other uses, such as parking, commercial development, traffic flows, and new green belts.

If an underground system is built, entire new landside development is created. An open-cut segment where the line runs just below the grade also provides the opportunity for new greening and urban design along the right-of-way.

A dynamic mix of housing and commercial use become feasible when zoning and land development are synchronized for maximum growth potential and economic benefit. Too often when a major new activity generator is planned within an existing city, the transportation system needs are overlooked in the initial proposal either by the private developer or by the public agency responsible for moving the new project forward quickly. The integration should occur from the beginning; the political mechanisms need to be strengthened to interject the transportation infrastructure possibilities at the earliest stages of program definition. Often it is too late after the project is built to retrofit for transportation.

Large-scale development needs to be flexible to the extent that it is capable of incremental growth. This is true for urban transportation systems, themselves under development, as well for the land use and major project development. New development densities should be evaluated in conjunction with availability of transit and accessibility to existing transit systems. Transportation as a zoning tool can encourage specific types of building and growth.

Public transportation systems have a direct impact on highway requirements. If highway demand can be lowered, we can reduce urban land set-asides for roads and parking. If transit systems can "intercept" along heavily trafficked corridors, the multimodal commute also reduces land-use needs as well as serving to relieve congestion in the urban core or central business district (CBD).

We used to think of quality of life as the opportunity to spread out in the suburbs and live a very private relaxed life at the end of a manageable commute, probably made in the family car. Today's definition of quality of life must be modified to add the diminution of congestion and stress during the trip to work, the opportunity to reach jobs now increasing in the outer ring of suburbs, and access to many educational and cultural centers. All this must be available within newly sensitive environmental constraints, and at a manageable cost.

Essential issues for the urban designer responsible for urban transportation infrastructure:

1. Long-range planning commitments
2. Coordination in planning at all levels: federal, state, county, local, and across all disciplines from housing to industry
3. Major creative financing mechanisms: public and public/private resources made available
4. Solutions to jurisdictional problems: a regional network perspective, overriding local authority
5. Meeting the technical challenges: problems inherent in introducing new technologies into older cities' systems and expanding existing services with minimal disruption

Regional transportation infrastructures are the vital networks that tie the inner and outer rings of a metropolitan area together and enable the daily interactions necessary for the growth and prosperity of an urban region.

## AVIATION INFRASTRUCTURE

### Role of Aviation Infrastructure

The aviation infrastructure is, relatively speaking, a young system of airspace networks and ground linkages, but due to its highly technological characteristics, is experiencing an obsolescence and a need for "third-generational" maintenance and upgrading, paralleling what we are finding today in the older urban infrastructure systems. Only 50 or 60 years ago, in the 1920s, flying was an adventure available only to the pioneers or the very wealthy. Even into the pre-jet 1950s, crossing the continent was a 10- to 12-hour trip, crossing the Atlantic, 18 to 20 hours. Today, coast-to-coast flying time is 5 hours, and the SST arrives in England only 3 to 4 hours of supersonic speed after takeoff in New York. In the United States in 1945, 4.3 million air passenger revenue miles were recorded for aviation. In 1976, 170 billion were recorded and in 1986, a decade later, the United States reached the level of over 370 billion air passenger revenue miles. Growing at the rate of over 10% per years, the 1990s air passenger revenue miles will exceed 500 billion (Air Carrier Industry, 1986, 1988).

National and international in services and scope, the aviation infrastructure must respond to worldwide issues of regulation, security, and safety standards, as well as to national aspirations, prevailing cultural values, and social desires. In the U.S. context, for example, the national characteristics of business competitiveness and technological interest can be observed in the planning and operation of the American aviation industry. We do not have a single national flagcarrier airline, and each urban community has developed an independent airport operating authority which competes for air traffic, passengers, airline home bases, and for their share of the annual national funding source. In addition, the critical industry issues are debated by all interested or affected groups, users, the community, labor, environmentalists, and so on. Solutions are proposed that try to respond across the needs spectrum. In England or France, by contrast, decisions on this and other infrastructure systems are reached in consultation with public and private decision makers, but the centralized public authority is not bound by these decisions, whereas in the United States there is a multifaceted decision process.

A duality exists; nations that compete for the air travel and air cargo markets must cooperate very closely to ensure safety and airway access worldwide. This fine-line duality includes maintaining a leadership role in aircraft technology and electronic surveillance systems, while protecting the necessary levels of military defense. The airport system must be a planned total system in which the most familiar component, the airport, is only the intermediate destination in a larger trip movement. It cannot be a point-to-point system since there are two distinct elements: the theoretical transport speed and the access time to the system.

As we go up the transportation hierarchy of advancing (newer) technologies, access time to the system increases:

| | |
|---|---|
| Walking | 0 time |
| Subway, local bus | 5–10 minutes |
| Average intercity travel (journey by motorway/highway) | 20–30 minutes |
| Airplane travel | $1\frac{1}{2}$ hours |

It is not uncommon today for a 1-hour shuttle air trip between two major cities to require almost 2 hours for access, a minimum of 1 hour at each end of the flight. This affects total trip convenience and travel cost and must be factored into the choice for this transportation infrastructure. In the last decade in this country, due to reduced fares brought about by a policy of deregulation, aviation has achieved new competitiveness with other traditional modes for short- and medium-haul trips: 90% of all intercity trips in 1985 in the United States were by air (Meyers and Oster, 1984).

Aviation is the image of twentieth-century mobility, a role that railroads played in the nineteenth century. As a young mode, aviation is expanding, with a growth potential that appears to be outdistancing the ability of the existing infrastructure to accommodate to that growth without significant constraints, at least for the short term. This growth translates into a potential for congestion to the system. In the United States today, for example, we have built very few new airports after Dallas–Fort Worth opened in 1973. Although according to the Federal Aviation Authority, several are on the drawing boards, it is quite likely that they will encounter unsurmountable cost and community obstacles before they can be realized. Instead, major hub airports, with their major commercial air carriers, are expanding and rebuilding in place to accommodate intensified demand at existing hubs, such as New York, Chicago, Atlanta, and San Francisco. In Los Angeles a recent thrust to create a neighboring hub at Palmdale, now a regional airport, is meeting extensive local opposition to its increased use, which may result in renewed efforts to expand the Los Angeles hub itself, apparently the more acceptable option.

Aviation plays a major role in a region's economic health. For a "headquarters" city with a major airline center, the on-airport staff can number 35,000 daily employees, about a third of which belong to the major carrier's base operations. Regional jobs associated with the airport, those dependent on it and those serving in related services, are estimated at between 200,000 and 250,000 (Port Authority of New York and New Jersey, 1984a,b, 1985a,b). In the late 1970s the New York metropolitan region was experiencing outmigration. When it began to reverse, in the 1980s, surveys were made by the Port Authority of New York and New Jersey to identify factors that encouraged economic growth. The availability of extensive domestic and international air services in the region were seen as a top priority for businesses considering relocation into the region.

Without doubt, the new technologies of aviation, plus telecommunications, unite the world in an unprecedented way, recalling the role played by the clipper ships in colonial expansion, and by the railroads in land development during the nineteenth century. Worldwide manufacture of products and product components became feasible when air cargo's reliability and cost competitiveness was established. Using Far Eastern and third-world labor sources for an expanding range of manufactured soft, perishable, and seasonal goods is the norm today, far beyond the traditional range of imported hard goods in the international markets.

The aviation system's infrastructure needs are often defined as requiring two major considerations for upgrading and expansion: (1) updating the air traffic control network, addressing the technological needs in the national air space; and (2) the airports, the land-based components with physical needs on the ground (Congressional Budget Office, 1979). To these two recognized areas of need we should add (3) the ground transportation linkages, without which the aviation system remains incomplete for point-to-point origin and destination travel.

## Regulation, Regulators, and Controls

In the United States, there are federal agencies that are major decision makers for the aviation system and that plan its future:

*Federal Aviation Administration (FAA):* regulates all civilian aviation, air commerce safety, airports, and the use of national airspace. Examples of this role are: setting of dimensional and performance standards for airport runways, and establishing flying height parameters.

*National Transportation Safety Board (NTSB):* participates in all safety-related reviews and investigations. At the federal level, is responsible for all transportation systems, not simply aviation.

*Transportation Research Board (TRB):* privately funded research-oriented group, organized into nationally designated committees to exchange technical and management information and to identify resources and new research needs for operating agencies and the aviation and other transportation industries.

*Civil Aeronautics Board (CAB):* sunsetted in 1978, but representing an important concept in the history of U.S. aviation. Regulation of the airline industry included the setting of all fares, and controlling routes and schedules. Deregulation, at the sunset of the CAB, was predicated to open up competition to many smaller specialized airline services and to allow the development of new markets by reducing fares. This did occur, but we are now seeing the reemergence of giant corporate entities, mergers of regional and smaller airlines.

*Air Transport Association (ATA):* privately owned air carriers are members of this privately sponsored association. It reflects industry views and practices and provides a forum for cooperation among airlines and for focused research to benefit air traffic operations.

*International Council of Airport Operators (ICAO):* addresses the issues of setting and maintaining international standards and practices. It also develops policy on a worldwide scale. Members can be national operators, the regional or local airport authority, a city agency, or a private operator.

## Components of the Aviation System

System distribution has three major elements: airspace, airports, and access (De Neufuille, 1976). In the United States, the airspace is controlled by the technology of *air traffic control* (ATC), a function of the FAA. Operating in several geographic ranges, *enroute traffic control centers* monitor and report on air traffic en route between regional control centers. *Approach control centers* function in a 25- to 50-mile radius of the airport, where there are often conflicts among several busy hubs using the same approach airspace. *Air traffic control towers*, situated on the airport itself, take over the guidance about 5 miles beyond the airport threshold and carry through with all airborne and on-ground airport control. *Flight service stations*, based at each airport, monitor navigational aids and safety along the airport runways and are often manned by former ATC staff, who frequently leave the high-tension control centers after a decade of the FAA tower environment. Control in the airspace is maintained over separate navigational air routes, defining altitudes and predetermined routes.

*Vector omni range* (VOR) is 12,000 to 18,000 feet and has a minimum width. *Jet routes* range from 18,000 to 45,000 feet in altitude. *RNAV* (Area Navigation) defines the same heights as jet routes, but aircraft can fly over a predetermined track without having to overfly ground-based navigational facilities, thereby allowing a more flexible use of airspace, which eases congestion on high-traffic routes. *FL 450* defines over 45,000-feet flights and is used for individual flights on a point-to-point basis.

At the airport there are three distinct physical components: airside, terminal, and landside. The airside is subject to FAA control for safety, monitored and directed by Air traffic control (ATC) in the tower. Additional navigational aids on the ground include ILS/MLS instrumentation, Instrument Landing Systems, approach lighting, visual clues, and signage. Specialized equipment for safety, such as bird distractors and weather indicators, also under FAA approval, may be placed on the airside.

Design parameters and standards for runways, taxiways, and the apron areas are established by the FAA, which publishes dimensional requirements upgraded periodically to accommodate the technological changes in aircraft. Domestic flights under 3000 miles can use runways shorter than the 500 feet minimum required for overseas long-range aircraft. As airport services expand, the runways must respond to changing demands. The dynamic geometry determining the layout and design of runways and taxiways is outside the scope of this chapter but is monitored and enforced by FAA review and investigation.

Included in an airport's airside infrastructure are "utility" systems for fuel supply and 400-Hertz power brought directly to aircraft parked at their assigned gates via underground dedicated lines. This underground service contributes to apron safety and speed in refueling and reduces congestion among service vehicles, but is expensive to retrofit at existing facilities. This installation is generally cost-effective only for the larger airports. Increasingly, airports are installing runway sensors which give visual and instrument clues for landing and docking guidance.

Significant airside activity occurs at maintenance facilities, hangars, and major maintenance bases for a single airline that concentrates its operations at a home port and can employ some 10,000 people in a 24-hour around-the-clock operation. Cargo activity is also linked to the airport's airside, with taxiway access to the main runways. At international airports a foreign trade zone (FTZ) is often established to encourage tax-free light manufacturing on site, and expanded storage capability, all adding to an airport's revenue base.

The airport terminal is more than a building type; it is, first, the modal interface for air passengers who are either changing from ground (local) to air (long-distance) modes and need circulation space at the curb frontage and at airport gates (Hart, 1985). The modal interface can also occur between interconnecting passenger distribution systems interlining between short-haul flights to longer journeys, or linking parking or public transit stations. The second function requires planners to consider the changing of movement type, since air passengers represent a continuing flow pattern and aircraft produce batch movements of up to 400 passengers per jet flight. This difference of movement type and scale requires the design of flexible staging areas in the terminal to accommodate both flow and batch movements. Third, the airport is a processing facility for ticketings, baggage handling, security checks, and customs clearance. Each function requires specially designed space, electronic technology, and inspection of administration control areas.

Airport access, the landside element in the infrastructure system, can be by private mode, auto or taxi, or by public mode of limo, bus, or rail. The public modes rarely attract passengers with family, baggage, or originating in suburban areas, where the density does not support frequent or inexpensive airport connections. Public mode access is more successful, in Europe for example, where the rail or transit system has been developed as an integral part of the airport. This occurs more easily when a single governmental entity has the responsibility for a multimodal system; policy and jurisdictional boundaries create a genuine interconnecting system.

In the United States today over 80% of air passengers using major hub airports choose their private auto for airport access, creating congestion problems on access highway networks, aggravated when airport peak activity coincides with other urban traffic peaks, such as when late Sunday returning air travelers share the surrounding arterials with drivers from seashore or mountain weekenders. In New York City, only 25% of peak-load highway traffic surrounding JFK is related to air travel (Port Authority of New York and New Jersey, 1987).

## Service Classifications

Classification by airport operations is one basic way to define the system:

*Commercial Aviation:* scheduled carriers, chartered carriers, and cargo carriers. Users pay costs to the system via ticket taxes. Pays the airport landing fees and service charges for fuel and power.

*General Aviation (GA):* air taxis, charters, private or corporate aircraft. Primarily unscheduled. Does not pay its own way: aviation fuel taxes are used as the source of funding. Pays airport landing fees, but these are not high enough to compensate for the specialized safety measures often required for the smaller GA aircraft using major hubs. This type of service is expected to grow, intensifying the funding problem and need for subsidation.

*Specialized Technologies:* include STOL (short takeoff and landing) aircraft or VOL (vertical takeoff and landing) aircraft, once the hope for congested urban areas where a shorter-runway or nonrunway field could be inserted into locations near the city core. Helicopters, valued for their time savings and similar nonrunway needs, continue to be perceived as unsafe, creating unacceptable noise within a built-up environment.

Classification by route structure uses the distance criteria.

*Long Haul:* nonstop, transcontinental or overseas routes, heavier aircraft with greater capacity, larger loads.

*Medium Haul:* generally, 300 to 1000 miles. This distance category expanded under U.S. deregulation, which spawned many smaller regional air carriers using hub-and-spoke route structures to improve load factors in their geographical territory.

*Short Haul:* commuter services such as the shuttle fly in very heavily trafficked corridors, where they compete in frequency and price with intercity bus, private auto, and rail. GA flights are frequently in this distance category.

Airport classifications can relate to level of air passenger activity and the role the airport plays in the overall system. They can also be identified by specialized service or aircraft.

*Major Hub Airports:* passenger activity is in the range of 15 to 20 million annual air passengers. Commercial aviation is the basis of operation, but GA and helicopter services also use the hub. Most likely to have a public transportation option available and experience congestion in airspace, airport, and on access highways. National and international services mean intense security, customs and cargo operations, foreign trade zone potential, and major airline carrier maintenance bases.

*Hub Airports:* smaller in size than their major counterparts in the system. Generally, provide a full range of services for commercial and GA flights. At least one public transportation access link will be available. Airline maintenance shops and headquarters may be located here, for medium-haul airline services.

*Reliever Airports:* located within close proximity to a major hub, these airports are looked upon as potentially able to accept some of the peak traffic from the hub. Within the same FAA air approach control center, requiring extreme skill in ATC. Usually exist in built-up areas that resist increases in environmental impacts.

*Regional Airports:* serving adjacent jurisdictions, such as Dallas–Fort Worth, where a large geographic region needs a concentration of air services to increase flights and cargo connections at a centralized hub. New regional airports are located as far as 30 miles from the cities they serve, to allow for expansion and to preserve extensive land barriers for environmental separation. Existing regional airports in areas under development or in built-up areas have the same constraints to growth as those of reliever airports.

*Military Airports:* often located in a historic defense network, or adjacent to military bases. Practice and testing operations may be part of their program. High security must be maintained.

*All Cargo:* an alternative to bringing cargo into major busy hubs, this specialized airport may benefit from less congestion and regulation than apply to major hubs, but must compete with the flexibility of using the excess belly capacity available in large passenger jets.

Ownership classifications can also be used to define the aviation infrastructure. Ownership operation by a central government, with airlines either privately or central governmentally operated: a "quasi" governmental authority or a public corporation established to run the airport; special public authorities operating one or a group of airports; department of state of a county, or a local transportation agency can be given the legal mandate to operate a local or regional airport. A private organization can also be in the airport operation business, but this occurs most frequently at smaller or more remote airports, or at publically owned facility that is privately operated under contract with the public agency.

## Aircraft Capacities

Aircraft capacities must be interrelated with airport compatibilities to accommodate a range of types and sizes of equipment. Airports must be flexible for the future aircraft to minimize operational difficulties in the terminals and on the runways. Aircraft produce batch movements with peak and stop cycles, producing very different impacts if a commuter plane brings fewer than 100 passengers or a jumbo jet dumps 400 people into the terminal at once. If future technology brings us a 600- to 800-passenger expanded jumbo plane, existing terminal compatibility will be difficult to maintain.

One severe runway problem already exists for major hub airports serving GA traffic in peak periods (Ashford et al., 1984). Runway clearance must be carefully adjusted in the wake vortex of jets, for the safety of smaller aircraft, which can otherwise be swept into the vortex of the jet takeoff. Policies to discourage GA use of these airports have been limited to raising GA fees, but the desirability of GA access for connecting to long-haul flights has overridden this attempt to segregate flights by aircraft size. Terminal design must accommodate the 150- or 400-person flight activity. Design and planning include providing the appropriate level of terminal services for customer processing, customs, baggage, food, and amenities. Terminal design must also have the flexibility dimensionally to adjust to demands of loading bridge apparatus for different aircraft, since the airport cannot afford technological obsolescence when faced with a changing demand for service.

## Constraints to the System

The aviation infrastructure has both new technology and operational strategies to apply in overcoming constraints in today's aviation systems, particularly those associated with rapid growth and peak demand. In addition to these, several other significant areas of constraint should be acknowledged.

Weather conditions create a level of uncertainty, increase the hazards of flying, and cause seasonal delays. Passengers can either accept the inherent delay, or select an alternative standby mode for short or medium trips. The aviation industry is developing and instituting high-sensitivity weather forecasting and more responsive runway sensors to alert to dangers from ice and snow. Safety standards have become more stringent, with national and international incidents spurring more careful inspections of passenger and cargo operations. Military priorities which have political overtones, can also become a system constraint, as can hijacking, drug traffic, and security.

Environmental constraints, particularly those that are noise related, have increased. Recent legislation in the United States mandates a 30% smaller noise footprint or mapped area of impact around airports, coupled with requisite retrofit of older aircraft engines to meet more rigid standards. By modifying patterns of approach and takeoff, noise impact is spread among different neighboring communities. Curfews for aircraft activity minimize noise disturbances in dense urban areas, although these time limits counteract the need to spread peaks out over a longer period to ease the operational congestion.

Airport air pollution is greater from ground vehicle activity and its congested peaks than is recorded from aircraft exhaust. Soil and water quality issues are related to oil seepage from both oil tank farms and underground lines. Airports must test and maintain, under the FAA and Environmental Protection Agency (EPA), an acceptable water quality level.

## Airport Terminal Design

Four basic terminal design concepts are in use today. A variety of combinations, utilizing the best of several alternative design approaches, meet changing needs and airport growth (Blankenship, 1974). One basic configuration may be adopted at the first stages of development, but to increase flexibility or to meet specialized physical environments, another may be grafted on as the airport develops.

The *pier* or finger concept requires the passenger to walk down one or more fingers extending out from the main terminal building. The advantage is an efficient use of airport space, but it requires aircraft tugs to push large aircraft away from the terminal in a power-in, push-out apron operation. Gate positions line each pier, with aircraft usually parked perpendicularly. This is a centralized processing concept, providing large passenger processing without using a great deal of land area.

The *satellite* concept has one or more satellite pods generally located away from the main terminal and connected either through an underground passageway or via an aboveground finger. Aircraft are parked nose-in, requiring a power-in/push-out operation. As distances increase, to accommodate more satellites and larger aircraft, a necessary and costly mechanical connecting link is required between the main terminal and the satellite. Some processing activities are decentralized, occurring in the main terminal or at the terminal frontages.

The *linear* concept, perhaps the most basic design, brings the auto or other ground transportation access modes as close as possible to the aircraft gates; walking distances are minimized. The attraction in using this design concept, the "gate arrival," is the relative ease of linear expansion, with duplication of these linear terminals along an endless spline as the airport becomes increasingly decentralized. Higher-density aircraft and the need for greater terminal security today make this concept relatively inefficient for major hubs.

The *transporter* concept, initially developed as a mobile lounge at Dulles Airport in Virginia, allows ground vehicles to operate from a compact central terminal to remotely parked aircraft. This allows the airport to construct a relatively small passenger terminal, but requires an ongoing operating expense for the transporter vehicles. In the early years of this concept, one vehicle could transport a full aircraft passenger load, but as aircraft capacities increased, more and larger vehicles were needed, adding to airside activity and expense. Both elevating and nonelevating transporters are used. This concept is in use today in peak seasons, adding remote "hard stand" gate positions on the apron and servicing them with buses or other specialized transporter vehicles called "planemates." The independence of aircraft operations and passenger terminal building operations make this a most flexible way to adjust to changing aircraft size and maneuvering requirements.

Vertical distribution concepts have developed to separate the flow of arriving and departing passengers. In the *one-level* concept, separation of passenger flows occurs horizontally. This is most suited to low-passenger-volume airports and is the most economical solution for circulation. *Two-level systems* can either separate passenger flows by level, or separate the passenger processing areas from the arrival baggage areas. When all passenger processing is done on the upper level, the passenger/plane interface is directly at aircraft doorsill height. When arrival and departure functions are on separate levels, landside vehicles can also be separated for a more desirable but more costly ground access accommodation. Variation in these basic designs allow for differences in passenger volumes, types of aircraft, and specific requirements of site and operation.

## Funding and Costs

To provide for the predicted growth of aviation in the United States, legislation in 1980 established the Airport Trust Fund, based on taxing the aviation industry on a pay-as-you-go basis: 8% tax on passenger tickets, 7-cent tax on GA fuel, $25 aircraft registration tax, 5% air freight shipping tax, and $3 tax per international passenger (Apogee Research Inc., 1987). This fund currently has a reported $4.8 billion in resources and is looked on with envy by public infrastructure advocates who operate in highly subsidized and needy sectors such as mass transit. A series of airport development (ADAP) legislative steps have allocated capital development funds to airport authorities, starting at 90% in 1976–1978, 80% in 1979–1980, and at 50% in the 1980s.

International airport ownership differs from that in the United States, where a federal centralized government provides the funds to local or regional public entities for capital development only, and the local operating agencies are responsible for operating revenues generated by airport real estate, parking fees, concessions, and private-sector airline landing and other fees. In Holland, and

Denmark, for example, a national airline, KLM, and SAS, is the flag carrier generating ticket revenues. The airports are under a single airport (or transport) cabinet member or department of the state, creating a more unilateral system and one that can most easily be coordinated with a national rail or ground transport connection for airline access. In the United States, historically, aviation represents a genuine public/private system in which the national government provides the standards and controls via the FAA, the capital investment through ADAP and Trust Fund resources, and allows private-sector airlines to operate relatively freely in a competitive market.

Current U.S. budgeting for infrastructure identifies the aviation system as needing 3% of the total infrastructure budget estimate of $53.4 billion. National highways and public transit are identified as needing 61.2%, or more than 20 times more in public funding resources (Congressional Budget Office, 1983). Aviation is different from public transit or rail, in that the airline passenger or user pays substantially for the cost of the system.

Delay has the highest economic cost to the aviation system; it is the most significant disincentive affecting the quality of the delivery of services. It can be critical in any part of the system. Passengers choose to avoid heavily used gateway airports, ATC information systems are overloaded, runway lineups take up to an hour to clear, and short-haul flights are held in place if destination landings are backed up.

With the sunset of CAB controls in the United States, deregulation increased aviation user choice, and through competition, opened up new markets, leaving the less profitable and smaller urban centers with diminished or nonexistent services and at a higher travel cost per ticket (Meyers and Oster, 1984). Airlines need a 55% load factor to break even. Fares are adjusted wherever possible to guarantee this level of use. Special discount fares are instituted but must be balanced by the ability to increase fares for service options or peak activity periods.

In 1987, a 21-hour day for 747 major airline aircraft cost $54.36 per minute, or $46,026 per day, to fly. Of this, labor represents 37% and fuel costs 25% (Anon., 1989). The smaller or newer airlines, using nonunion or low-scale wages, strive to lower the labor factor by concentrating on special markets and special fares.

In the 1980s, the latest 747 or 767 jet with 300 to 400 seats cost $41 to $101 million. For the 727 and 737 aircraft with 150 seats, the estimated initial cost is between $25 and $35 million. The FAA operating budget, in general terms, receives 75% of its operating expenses from passenger and aircraft operations, and 100% of its capital expenses fully reimbursed via user charges, the ticket tax.

Airport operating costs depend on landing fees and charges for fuel, power, and other services to aircraft, parking fees, concession contracts, real estate leases, and utility distribution charges. Their expenses reflect the 24-hour operation of a small city, emergency staff and services, maintenance responsibility, safety provisions, inspections, labor, and new construction not funded by the federal trust fund.

## Planning for a Growth Infrastructure

The development of a dynamic and responsive aviation infrastructure depends on the planner's ability to match system capacity with demand and to match aircraft fleets with seat demand (De Neufuille, 1976). This in turn is dependent on the ability to forecast this demand far enough in advance to order equipment and build the public facilities. In fact, aviation is considered a high-risk industry since commitments must be made several years ahead on the basis of a series of assumptions that are not necessarily going to hold up under changing economic conditions.

To complicate the forecasting methods for the aviation industry, long-term demand should be defined in terms of specialized markets, targeting the demand profile to these various market segments. Full-service airlines provide a traditional business class with baggage and food service; "no frills" and charters charge a minimal base fee and provide no services. Business and personal travel have unique demands.

As with other public utilities and services, aviation's ability to service the peaks, or to spread peaking out more equitably in the system, and to control overexpansion, is critical in both long-range planning and short-term economics. Ideally, planning should be concentrated on the short term, with frequent cross checking in technology, operations, and passenger and cargo activity. But the need to plan for an orderly long-range system remains.

The peaks in the system create the most costly delays. An airport's ability to spread these surges of activity are limited by the time constraints of typical domestic business days and seasonal or holiday surges. Except in select high-density corridors served by shuttle on-demand flights, the peak air traffic demand for scheduled flights is also constrained by available air navigational routes and airport gate slots at busy hubs. The system and its users have been forced to accept frequent daily, hourly, and seasonal peaks, but the increasing number of hours of high-peak conditions at major airports is what indicates a crisis level of growth and activity. Attempts to flatten the peaks include adjusting landing fees to favor off-peaks or edges of peaks. Lower fares are offered on off-peak days, seasons, or departure times. These strategies are only partially successful, since the realities of airport curfews and business day schedules, plus the European "window" that dictates early evening departures for morning overseas arrivals, perpetuate congestion and delay.

In addition to planning for the quantitative needs by forecasting and the qualitative needs associated with congestion and delay, the physical needs of today's expanding airport systems require long-range land-use planning to preserve, wherever possible, sufficient vacant land as a buffer zone. As aviation activity increases to meet the demand of a growing industry, the impact of that activity becomes more burdensome to neighboring communities. Where it is still possible to land-bank for the future, airports should be in the vanguard of rezoning for sufficient space to allow for compatible uses and appropriate buffers. In the technological arena, flexibility in the physical design of aviation facilities must be understood in relation to design changes in aeronautics and avionics, recognizing that just as technical obsolescence cannot be accepted for aircraft and air traffic control infrastructures, airports must have the capacity to remain technically current and responsive to the physical demands of the system's other components.

## Urban Design Issues and Impacts

The modern airport is itself a single-industry town, a specialized city that never sleeps. Active 24 hours a day, programmed and equipped to respond instantly to any type of emergency from weather to political explosions, it is a major activity generator as well as a magnet for regional growth. Many of the major airports in U.S. metropolitan areas were originally privately owned airstrips or military installations on the city outskirts, adjacent to open waterways, and 10 or 15 miles from the city core. Frequently, these sites were considered unsuitable for other uses and were therefore made available for airports. As the city grew, demand for air service increased and a local authority took over development and operation. Unlike other major activity centers, which today are located only after extensive market research and demographic analysis, most older airports were in place before the air passenger growth of 1950–1970. Airports scrambled to keep up with demand, and today are experiencing significant constraints to growth if located within built-up urban areas.

The desirability for airports and land-use compatibility is illustrated by the locating of new airports to serve growing regions at great distances, up to 30 miles perhaps, from the cities they are to serve (Hornjeff and McKelvey, 1983). This will provide an adequate opportunity to develop compatible land uses in the ample buffer zone. For older, more densely urbanized locations, however, no such land luxury exists. Immediately adjacent to some of our older airports, some kind of light industrial zone operates, usually businesses that have direct relationships to on-airport services such as cargo and food catering. At one time this property was not in danger of becoming residential, but as more desirable suburbs become too expensive or too far out, developments began to cluster under the path of aircraft approaching these busy airports. It is this clash of incompatibility between a noisy and somewhat dangerous activity center that creates congestion on the surrounding access highway network, and the demand for quiet residential neighborhoods, that is causing opposition. Where a smaller airport has tried to expand in the outer ring of suburbs, the hue and cry of residents is even stronger, since the residential quality has been established far longer than has the density level that now creates the demand for increased air traffic. When setting the site selection criteria today that will include the factor of distance, the concern for accessing the distant airport must also be put on the drawing board, so that ground access becomes an integral part of the design program.

Jurisdictional coordination, between neighboring communities and across county and state boundaries, is an essential urban design framework for strengthening the aviation infrastructure of air and ground components in the system. In defining the aviation infrastructure as airside, airport, and access, the urban designer should also consider the airport itself as the symbol of the system, and as such, recognize its role as icon and magic gateway to the city. Often this gateway serves as a symbol of the country or the community, glorious to behold from the air, yet often clogged and unwelcoming at the point of entry; the symbol and anticipation of the urban experience beyond is somehow diminished. The airport's very scale, the endless corridors for security and segregation of passenger flows, can be intimidating. The psychology of uncertainty, plus the physical stress of long-distance travel and jet lag, is an important programmatic input for designers of airports. In addition, there may be confusion in finding the right connection to take you from the airport point of entry to your actual destination.

Unique in its high technology, in its "invisible guideways" of airspace, yet remaining dependent on other urban infrastructures, earthbound at both ends, the aviation infrastructure must be understood to encompass the entire journey.

## REFERENCES

Air Carrier Industry, *Traffic Statistics Quarterly*, Transportation Systems Center, U.S. Department of Transportation, Research and Special Programs, Cambridge, Mass., 1986, 1988.

American Public Transit Association, *Transit Capital Needs, 1984–1988*, APTA, Washington, D.C., 1984.

Anon., A Day in the Life of a Jet, *The New York Times, Sunday Magazine*, April 1989.

Apogee Research, Inc., *The Nation's Public Works: The Report on Airports and Airways*, National Council on Public Works Improvement, Washington, D.C., 1987.

Ashford, Norman, H. P. Martin Stanton, and C. A. Moore, *Airport Operations*, Wiley, New York, 1984.

Blankenship, Edward G., *The Airport*, Praeger, New York, 1974.

Congressional Budget Office, *Public Works Infrastructure: Policy Considerations for the 1980s*, U.S. Government Printing Office, Washington, D.C., 1983.

Congressional Budget Office, *Public Works Infrastructure*, U.S. Government Printing Office, Washington, D.C., 1982.

Cooper-Hewitt Museum, *Subways of World Examined*, The Museum, New York, 1977.

De Neufuille, Richard, *Airport Systems Planning*, MIT Press, Cambridge, Mass., 1976.

Georgano, G. N., ed., *Transportation through the Ages*, J. M. Dent, McGraw Hill, N.Y., 1972.

Hart, Walter, *The Airport Passenger Terminal*, Wiley, New York, 1985.

Hornjeff, Robert, and F. X. McKelvey, *Planning and Design of Airports*, 3rd ed., McGraw-Hill, New York, 1983.

Meyers, John R., and C. V. Oster, Jr., *Deregulation and the New Airline Entrepreneurs*, MIT Press, Cambridge, Mass., 1984.

Port Authority of New York and New Jersey, *Annual Report*, Port Authority, New York, 1984a, 1985a, 1987.

Port Authority of New York and New Jersey, *Aviation Department Annual Reports*, Port Authority, New York, 1984b, 1985b.

Pushkarev, Boris S., and Jeffrey Zupan, *Urban Rail in America*, Indiana University Press, Bloomington, Ind., 1980.

Ruppethal, Karl M., ed., *Energy for Transportation*, Center for Transportation Studies, Vancouver, British Columbia, Canada, 1981.

Urban Land Institute, *Policy Forum in Joint Development of Rail Transit Facilities*, ULI, Washington, D.C., 1987.

Urban Institute, *The Nation's Public Works: Report on Mass Transit*, National Council on Public Works Improvement, Washington, D.C., 1987.

Vulchic, Vukan R., *Urban Public Transportation*, Prentice-Hall, Englewood Cliffs, N.J., 1981.

# ADDITIONAL READING

## Rail/Transit Infrastructure

Institute for Mechanical Engineering, *Rapid Transit Vehicles for City Services*, Symposium Proceedings, IME, 1969–1970.

Olmsted, Robert P., *The Diesel Years*, Golden West Books, San Marino, Calif., 1975.

Port Authority of New York and New Jersey, *Annual Report,* Port Authority, New York, 1985, 1986.

Pushkarev, Boris S., and Jeffrey Zupan, *Public Transportation and Land Use Policy,* Indiana University Press, Bloomington, Ind., 1977; and *Urban Rail in America,* Indiana University Press, Bloomington, Ind., 1980.

Richards, Brian, *New Movement in Cities,* Reinhold, New York, 1966; and *Moving in Cities,* Westview Press, Boulder, Colo., 1976.

Sampson, Roger J., and Martin T. Farris, *Domestic Transportation,* 2nd ed., Houghton Mifflin, Boston, 1971.

### Aviation Infrastructure

Air Carrier Industry, *Traffic Statistics Quarterly,* Transportation Systems Center, U.S. Department of Transportation, Research and Special Programs, Cambridge, Mass., 1986, 1988.

International Air Transport Association, *World Air Transport Statistics,* IATA, Geneva, Switzerland, 1987.

Port Authority of New York and New Jersey, *Annual Reports, 1983–1988,* and *Aviation Department Annual Reports, 1985, 1986, 1987,* Port Authority, New York, 1983–1988.

## QUESTIONS

1. Compare and contrast the roles of aviation and rail infrastructures in the transportation network.

2. Discuss energy savings in rail systems. Compare transit to auto use.

3. Why must public transit in the United States be subsidized? What factors could lower the degree of subsidation needed?

4. Identify the relationship between density and inner-city transit or suburban rail service levels.

5. What three major characteristics of public transportation alternatives determine a passenger's modal choice for journey to work trips?

6. Name five types of classifications that can be used to define a public transit system.

7. Discuss the relationship between land use, land values, and the growth of urban/suburban public transportation infrastructures.

8. What is the average life span of a heavy rail subway car? How would you extend this life?

9. Contrast the planning and design issues facing the designer of an inner-city transit line extension with those issues to be addressed for a suburban or intercity line expansion.

10. What are the opportunities for design and economic growth when building (a) an underground transit line; (b) an overhead line; (c) a cut-and-cover service?

11. How would you determine whether a light rail, subway, or public bus system would be the most suitable system for a newly developing county and its dense urban core?

12. In locating a new airport for a rapidly developing urban center, name at least five planning considerations to be addressed.

13. Select the optimal location for a new airport to serve a city of 2,000,000, or more than one city of over 1,000,000 each, in these geographic locations:

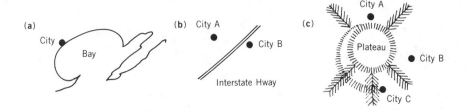

14. What are the specific challenges for a highly technological infrastructure such as aviation?

15. To ensure aircraft–airport compatibility, what design components should be examined?

16. Define *batch* movements. What problems do they create in airport design and operation?

17. List the physical components of the airside of an airport.

18. What types of on-site infrastructures do airports provide on (a) the landside of the terminal and airport; (b) the airside of the airport?

19. Who pays for the aviation infrastructure in the United States?

**20.** List the basic configurations used in airport terminal design and describe the advantages of each.

**21.** How is an aircraft guided through the airspace from one hub airport to the next?

Port Authority Bus Terminal

# 10

# BUSES

**Nicholas Bellizzi**
The Hudson Partnership Inc.

## INTRODUCTION

Buses are a component of the overall transportation mix, which includes rail systems (subways, light rail, commuter rail, freight, etc.), automobiles, trucks, taxis, limousines, ferries, airplanes, and helicopters. Bus systems are the most commonly used component of urban mass transportation systems. Buses provide important mobility within the cities of the United States. Each year they serve some 1.5 billion person-miles of travel. Nationally, buses carry more than 70% of all public transit rides. Excluding New York City, they serve more than 85% of all public transportation passengers nationwide. Buses are readily available and easily purchased; are flexible in terms of routes traversed; utilize existing streets and highways, thus minimizing capital expenditures; are manufactured in many sizes and capacities; and utilize various forms of fuel (gas, diesel fuel, electric, propane, methane, etc.). Bus systems provide both intra- and intercity travel needs, transport large volumes of students to and from school each day, provide charter service to major events (theater, ball games, parades and exhibitions, fairs, etc.), and serve retail, office, and residential areas with both local and express service.

While rail transit systems efficiently serve high-volume densities of travel, bus transit system serve less densely traveled corridors and areas. Routes are spaced to keep walking distances within $\frac{1}{4}$ mile, and stops are located relatively close together. While local bus travel is slow compared to automobile or rail modes, express bus and intercity bus travel with limited stops and line-haul operation is suitable for longer bus trips. Exclusive bus lanes, high-occupancy vehicle (HOV) lanes, and preferential bus treatment can also be utilized to reduce travel times for all types of bus transit.

Although a typical transit bus is at least 25 feet long and propelled by a diesel engine, many new and innovative vehicles are used for bus transit. These vehicles include double-deck and articulated buses, 14- and 20-seat passenger vans, buses specially outfitted for the transport of handicapped passengers, and other such specially designed vehicles.

## THE ROLE OF BUS TRANSPORTATION IN THE TRANSPORTATION MIX

Urban mass transit systems are generally oriented to serving work, school trips, and regularly scheduled service to large activity centers such as shopping districts, recreational areas (stadiums, beaches, etc.), and central business districts (CBDs). The purpose of increasing and maintaining bus ridership are:

1. Increased ridership brings in more revenue for transit operations.
2. It serves as a primary commuter service in reducing automobile traffic and associated congestion and traffic jams.
3. In encourages traveling to businesses, entertainment, and services within the community, all of which need community support and are in competition with nearby communities.
4. It gives time–cost–convenience advantages to the public vehicle rider because it is less expensive, often faster, and frequently as convenient as or more convenient than automobile travel.
5. Decreases in automobile usage aid in the attainment of air quality standards in urban areas.
6. Special segments of the market, such as the young, elderly, handicapped, poor, and nondrivers, are transit dependent, and the mobility of these market segments depends solely on public transit.

## SERVICE TYPES

The product planning function involves the types of service, the quality of that service, and access afforded to potential riders. Bus transit may be divided into three major categories:

1. *Regular Routes:* configuration of the routes and schedules operated, which are the primary variables.
2. *Special Services:* those services not following regular routes, such as services to sporting events.
3. *Charter Service:* private service packages under contract to a special group, such as junkets to entertainment centers.

Service on the major route categories described above may be one of the following types:

1. *Local:* serves all stops along a route. This is the most basic and common form of urban bus service.
2. *Express:* more of a line-haul operation, with limited pickup and drop-off stops on either end of an express run, usually running line-haul along a highway or major arterial roadway.
3. *Shuttle:* connects two points or activity centers with few or no stops in between.

4. *Intercity:* terminal-to-terminal operation between cities.

5. *Suburban Commuter:* multiple pickup and discharge points in bedroom communities running line-haul into a bus terminal in the CBD (e.g., Port Authority's West Side Terminal in Manhattan).

6. *Secondary Distribution:* includes routes for internal service within major activity centers. These routes may provide connections between such centers and major parking facilities and/or transportation terminals. In airports, such systems provide transit to long-term parking areas, nearby train and bus terminals, car rental centers, and other activities integrated within the airport area.

7. *Subscription:* passengers are picked up at a single location and transported to a single destination en mass.

8. *School Bus:* transport children to/from various grade levels in specially designated yellow school buses.


## VEHICLE CHARACTERISTICS

### Standard Vehicles

There are four basic types of bus in general use today (Polytechnic Institute of Brooklyn, 1971):

1. *Transit Bus:* a two-door vehicle with automatic transmission, low-back seats arranged in a transverse–parallel mixed pattern, an automatic front door, and generally powered by a diesel or gasoline engine.

2. *Suburban Bus:* a two-door vehicle with standard transmission, a mixture of high- and/or low-back seats, generally in the transverse pattern, automatic front door, generally powered by a diesel engine.

3. *Intercity Bus:* a single-door vehicle with standard transmission, high-back seats in the transverse arrangement, manual front door, and luggage compartment below the floor; powered by a diesel engine.

4. *School Bus:* an inexpensive gasoline-driven vehicle with low-back seats, standard transmission, and a manual door.

There is a trend toward air conditioning each of the first three vehicle categories. The *suburban bus* is a hybrid vehicle with characteristics between those of the transit bus, designed for maximum capacity and heavy short-distance use, and the *intercity bus*, designed for maximum passenger comfort and long-distance usage. The major manufacturer of transit buses in the United States is the General Motors Truck and Coach Division of General Motors

Corporation (GM). General Motors, for example, has produced about 90% of all transit buses used in the New York metropolitan region. The second manufacturer of note is the Flxible Company.

## Special-Purpose Vehicles

In addition to the standard transit, suburban, and intercity buses, a number of special-purpose vehicles have appeared on the market. The minibus is one such vehicle. The term *minibus* is derived from its original major manufacturer, Minibus Inc., but is commonly used to describe small buses with a capacity of 18 to 25 seats used for short-distance, low-volume transportation and circulation. These vehicles range from converted vans such as the Ford Econoline, to the buses manufactured by Minibus Inc. and the Flxette model by Flxible.

Although not in wide use in the United States in the past, articulated buses have been widely used in Germany, Austria, China, and Italy. The articulated bus is essentially a bus with an integrally connected trailer unit, connected by a joint capable of bending in the horizontal and/or vertical planes. The advantage of such vehicles is the approximately 40% increase in carrying capacity over a conventional bus without corresponding rises in operating costs, particularly labor. The bus is operated by a single driver. Articulated coaches were built and introduced in the United States as early as 1938. Their use, however, was not widespread; only 20 such vehicles were purchased between 1938 and 1950.

Despite the apparent ability of the articulated bus to negotiate city streets, turning sharp corners and making frequent curb stops cause some operating difficulties. It appears, however, that the use of such vehicles for express routes or for bus rapid transit is promising. On such routes, turning and stopping are infrequent. Cities and suburban communities such as the New York metropolitan area and Westchester County currently utilize articulated buses on several routes.

## PERFORMANCE CRITERIA

The ranges of weights of bus transit vehicles are shown in Table 10-1. Typical vehicle relocation and stop spacing are shown in Table 10-2.

## SERVICE STANDARDS

Service standards are recognized as one of several policy statements that can guide transit decisions. For example, design standards relate to the dimensions of physical facilities (e.g., bus depots). Performance standards measure the

**TABLE 10-1   RANGE OF WEIGHTS OF BUS TRANSIT VEHICLES**

| Transit Vehicle Type | Weight (Empty)<br>(lb × 1000) | Design Capacity<br>(lb/passenger) |
|---|---|---|
| Van | 5–7.5 | 200–650 |
| Minibus | 7–17 | 200–750 |
| Transit bus | | |
|   Single unit | 14–26 | 175–340 |
|   Articulated | 28–36 | 160–360 |
|   Double deck | 15–28 | 200–230 |

Source: Institute of Transportation Engineers (1982), p. 185.

efficiency and effectiveness of transit resources (e.g., buses per mechanic). These indices assess how well transit managers utilize their resources.

In contrast, service standards focus on the decision of when, where, and how bus service should be operated. Service standard policy is used to develop a bus system's service plan and monitor results. Service standards policy provides an analytical framework for monitoring service on an ongoing basis. The shift in ownership in bus systems across the county, from private ownership to public ownership or to government-assisted private ownership, created a problem in the evaluation of bus service. Revenue generation and profit are no longer primary determinants in gauging performance. Policy guidelines now cover the quantity of service as well as minimum performance levels related to service quality. A list of 15 separate service standards within four broad service categories are presented in Table 10-3.

**TABLE 10-2   VEHICLE RELOCATION AND STOP SPACING**

| Transit<br>Vehicle and<br>Service Type | Maximum<br>Performance<br>Speed<br>(mph) | Platform<br>Speed<br>(mph) | Linear Stop Spacing | | |
|---|---|---|---|---|---|
| | | | CBDs<br>(ft) | Traditional<br>Practice<br>(ft) | Same Modern<br>System with<br>Longer Stop<br>Spacings<br>(ft) |
| Urban bus, local | 50–65 | 8–14 | 500–1000 | 500–800 | 1000–1500 |
| Limited bus | 50–65 | 12–18 | 500–1000 | 1200–3000 | 2000–15,000 |
| Express | 50–65 | 16–32 | a | 4000–30,000 | 5000–15,000 |

Source: Institute of Transportation Engineers, (1982), p. 184.
aUsually stops at limited number of locations in the CBD.

**TABLE 10-3  SERVICE STANDARDS**

I.  Service coverage
    1. Availability
    2. Frequency
    3. Span
    4. Directness
II.  Patron convenience
    5. Speed
    6. Loading
    7. Bus stop spacing
    8. Dependability
III.  Fiscal condition
    9. Fare structure
    10. Farebox recovery
    11. Productivity
IV.  Passenger comfort
    12. Waiting signal
    13. Bus stop signs
    14. Revenue equipment
    15. Public information

# PLANNING CONSIDERATIONS

Several fundamental considerations and assumptions provide the foundation for bus system service programs. These factors are outlined below.

## Street and Roadway Conditions

The effectiveness of a bus service is a function of many internal and external factors. Internal factors are generally within the control of management and include route and schedule development. External factors, on the other hand, are beyond management's control and include the character and density of development as well as the socioeconomic characteristics of patrons served.

One of the most important external factors affecting the effectiveness of public transit are street and roadway conditions. Narrow congested streets, parking, illegal and double parking, traffic signal system timings, deteriorated roadway pavement, flooding, and other such environmental and infrastructure factors all contribute significantly to overall speeds and system effectiveness. Cost per mile and speed are inversely related.

## Service Utilization

Service utilization can be measured in several ways, including are review of productivity and passenger loading. Productivity, or the number of passengers generated for each hour of service operated, is a basic measure of service effectiveness. Productivity is, in turn, related to internal and external factors. The numerator—passengers carried—is related to external factors such as the characteristics of the communities served.

## Network Geometry

Very few bus transit systems in the United States have been "master planned." Most systems evolved over time in stages, giving rise to overlapping, duplicate, and discontinuous service.

## Bus Priority Treatments

One of the disadvantages of the use of local buses is the fact that they are subject to the same traffic delays and hazards as are private autos. In congested downtown areas, this becomes a considerable drawback, as average travel speeds are extremely low. A number of measures may be taken to expedite the flow of vehicles to and through downtown areas, including preferential treatments such as exclusive lanes for buses and preferential control procedures.

Bus priority treatments vary in design, planning, and operating concepts. Bus priority treatment can represent contraflow operations, queue bypasses, exclusive use of freeway or roadway lanes or entrance/exit ramps, traffic signal preemption, special turn permission or signal phases, special use of streets (bus-only streets), bus lane bypass of toll plazas, busways in railroad or freeway medians right-of-way, and so on.

As most major cities around the world have given increased attention to "traffic management" programs to reduce auto dependency and associated congestion/air quality problems, greater attention has been given to preferential treatment of high-occupancy vehicles (HOVs). HOVs include buses and carpool/vanpool vehicles generally with three or more passengers.

# BUS STOPS

Bus vehicle and overall system performance is greatly influenced by the location, design, spacing, and operation of bus stops, including terminal location and design. Bus stop spacing is a primary determinant of operating speeds and affects overall route travel times and patron walking distances.

Bus stops are generally located adjacent to curbs on streets, thus providing for patron access via sidewalks. Signs and street and curb markings designate bus stops. Pole-mounted signs indicate the bus route number and can provide route, schedule, and other pertinent information to waiting passengers. Benches and bus stop shelters may also be provided, especially at busy stops or where bus headways are infrequent.

Bus stop locations can be in one of three locations: "near side" or "far side" at intersection locations or at midblock locations. The selection of the bus stop location is made on the basis of curb space requirements, intersection turning movement volumes, transfers to nearby or intersecting bus routes, and the location of major pedestrian generators and activity centers.

## FARE STRUCTURE

As in the case of rail rapid transit, buses may use a flat fare or a zone fare. All fare collection, however, is on board the bus and is monitored by the driver. In addition to driving the bus and collecting the fares, the bus driver may be required to make change and answer questions. This zonal fare collection represents a considerable problem.

Virtually all local services utilize a fixed fare. In past years, there has been a trend toward exact fare systems, which operate with locked fare boxes, in which the bus drivers cannot make change. Such fare systems were initially developed to discourage robberies, but speedier boarding of passengers has also resulted. It is important, however, that fares be an even amount with such exact-fare systems.

Zone fares are most often used on express or suburban lines. They provide for the easy monitoring of fares. The assumption is that all trips either originate or end in the CBD area. Thus on inbound trips, full fare is paid on vehicle entry—the fare depends solely on the stop at which the bus is boarded. On outbound trips, a flat fare is paid on entry, the remainder paid when exiting, depending solely on the exiting stop.

## COST OF SERVICE

### Economics

A standard model transit bus (GM) costs approximately $135,000. The useful life of a transit bus can be expected to be in the range of 15 to 25 years, depending on use, maintenance procedures, and geographical area of operation. Operating costs are the overwhelming concern of bus operators, as capital costs are relatively small by comparison. Labor is the chief element of operating costs, as each bus requires an operator, in addition to supervisory, maintenance, and other personnel. There is a substantial difference in the total operating costs reported in different parts of the country, such as in San Francisco versus New York. Past studies of reported cost for local bus operators indicates a relationship between the population density of the service area and the cost per bus-mile (Polytechnic Institute of Brooklyn, 1971).

## Farebox Recovery

Bus transit services must obviously operate within a budget. Within this budget, the first objective is to provide a system of transit services that will offer the best possible service to residents. To accomplish this, each is examined individually to determine if any bus line is placing an inordinate financial burden on the entire system. Routes are compared to systemwide averages periodically so that operating deficits are controlled and equipment is deployed in a productive manner. As part of this standard, two measures are generally used. First, each private bus operator should recover overall a minimum of 55% of operating costs from the farebox for local service operations. Second, individual route performance, as measured by the farebox recovery percent, should be monitored at least once per year. The standard for individual route performance defines suggested actions if certain levels are not achieved, as shown below (NYC Department of Transportation, 1987):

| Relative to Systemwide Standard or Actual System Average (%) | Suggested Actions |
|---|---|
| Above $66\frac{2}{3}$ | Acceptable, modify as required |
| $50$–$66\frac{2}{3}$ | Review, modify service |
| Below 50 | Unacceptable, consider major change or elimination |

## Bus Allocation Model

The basic concept underlying the cost allocation model is that each expense account is influenced by one or more resource levels. Consideration of the nature of various operating expenses has resulted in the identification of three major resource levels that "drive" each particular expense item: (1) vehicle hours, (2) vehicle miles, and (3) peak vehicle requirements. The use of these operating statistics is generally consistent with the data collection procedures found in the UMTA Section 15 reporting requirements employed by transit operators. One such model is a *three-variable model*. A three-variable model is often used on the basis that it is easier to develop and apply while maintaining a degree of accuracy comparable to more complex formulas. It should be recognized that the model approach selected in such an analysis should be consistent with the study objectives to analyze bus routes. Other models that distinguish more completely between the cost of providing peak and off-peak service could be applied where comparisons between service types (e.g., local versus express) are to be made.

Typical cost models developed are termed a *fully allocated cost formula*. This method receives its name from the fact that all operating costs are included in the model development. As a result, the sum of the individual route costs produced through use of the model will equal the total operating costs of each carrier (NYC Department of Transportation, 1987).

## Operating Expenses

The primary data source in the cost allocation model for each operator is the operating expenses for a one-year period. The selection of this 12-month period provides a recent picture of operating expenses. Consistent with the present financial reporting system, all expenses are tabulated in accordance with the federal Section 15 reporting system.

The UMTA Section 15 accounting structure provides a three-way classification of expenses: function, object, and mode. Function describes the various types of activities (e.g., vehicle operations), while the object classes describe general categories of expense (e.g., wages and salaries). All subsidized transit operators are required by the Urban Mass Transportation Administration (UMTA) to follow the Level R reporting system. With this chart of accounts, there are four major function categories, which are further divided into 14 object subaccounts. In addition, there are other object subaccounts, listed under Reconciling Items (i.e., interest expenses, depreciation, etc.). Object classes consist of 10 major categories which are divided into subaccounts. With this reporting system, each expenditure is identified as both a function and an object class identification. Although numerous accounts exist, only those having dollar values are used in the analysis to develop the cost model (NYC Department of Transportation, 1987).

## Model Calibration

In the preceding section we provided an overview of the approach used in the cost analysis. The next step is to quantify the cost allocation model for each operator based on recent operating and cost experience. The cost formula is calibrated by performing the following three tasks:

## TABLE 10-4  ALLOCATION OF EXPENSES

| | Basis for Allocation (%) | | |
|---|---|---|---|
| Expense Account | Vehicle Hours | Vehicle Miles | Peak Vehicles |
| Vehicle operations | | | |
|   Operators' salaries and wages | 100 | | |
|   Other salaries and wages | 100 | | |
|   Fringe benefits | 100 | | |
|   Services | 100 | | |
|   Fuel and lubricants | | 100 | |
|   Tires and tubes | | 100 | |
|   Other materials and supplies | | 100 | |
|   Taxes | | 100 | |
| Vehicle maintenance | | | |
|   Other salaries and wages | | 100 | |
|   Fringe benefits | | 100 | |
|   Services | | 100 | |
|   Other materials and supplies | | 100 | |
|   Utilities | | | 100 |
|   Taxes | | 100 | |
| Nonvehicle maintenance | | | |
|   Other salaries and wages | | | 100 |
|   Fringe benefits | | | 100 |
|   Services | | | 100 |
|   Other materials and supplies | | | 100 |
| General administration | | | |
|   Other salaries and wages | | | 100 |
|   Fringe benefits | | | 100 |
|   Services | | | 100 |
|   Other materials and supplies | | | 100 |
|   Utilities | | | 100 |
|   Casualty and liability costs | | 100 | |
|   Taxes | | | 100 |
|   Miscellaneous expense | | | 100 |
| Other | | | |
|   Interest expenses | | | 100 |
|   Leases and rentals | | | 100 |
|   Depreciation | | | 100 |
|   Miscellaneous | | | 100 |

Source: NYC Department of Transportation (1987).

1. Assign each individual expense in the system financial statement to one of the three selected resources that influence costs.
2. Sum the costs assigned to each resource to obtain the overall cost allocated to the resource.
3. Divide the overall resource cost by the quantity of the resource used by the system. These calculations produce the unit cost of each resource that is a coefficient of the cost model.

The allocation of each item is made on the basis of judgment, although the relationship between the expense item and variable is typically quite evident. This is especially true for major expenditure items that comprise more than 50% of all operating expenses. As shown in the example on Table 10-4, each individual expense account is allocated to a particular resource.

Expense account categories are described below.

1. *Vehicle Hours.* Wages paid to drivers is the single largest expenditure. When combined with other payroll costs, such as fringe benefits, the importance of drivers in establishing the cost structure is evident. Employees used to operate buses are paid on an hourly basis. For this reason, the wage and fringe benefit expenses for drivers are allocated to vehicle hours. Supervision of operations is directly related to the hours of service provided, and this account has been allocated to vehicle hours.
2. *Vehicle Miles.* Many costs are directly related to the miles of operation. Expenses such as fuel, tires and tubes, and maintenance of revenue equipment are a direct function of the number of miles operated. In addition, insurance and liability expenses are a function of accident exposure in terms of miles of service.
3. *Peak Vehicle Requirements.* Many expense accounts do not vary as a function of either vehicle hours or vehicle miles. For example, the cost resulting from providing vehicle storage is determined by the maximum number of vehicles in service. Similarly, expenses related to maintenance of buildings, garages, service equipment, and shop expenses are influenced by peak vehicle requirements. A number of general and administrative expenses also vary with peak vehicle requirements.

## PRIVATIZATION

Cost savings can be achieved through the privitization of public transit systems through private consulting. For example, the City of Phoenix (Phoenix Transit), the Peninsula Transit District Commission (PenTran) in Virginia, and the Southeastern Michigan Transportation Authority (SEMTA) all currently contract out portions of their bus systems. Just as private financing for toll road construction is replacing the government's financing of Interstate highways in many states, private-sector initiatives are competing with or replacing public bus systems.

In New York City, midtown and lower Manhattan are the focal points of peak-hour commuter van and express bus services from the other four New York City boroughs, as well as other New York counties, New Jersey, and Connecticut. Within this market area, a great deal of private bus service flourishes. For example, private interstate express bus service is being provided without requiring a franchise from New York City. Interstate express buses need only carry Interstate Commerce Commission (ICC) certificates. Similarly, smaller vehicles, such as vans, obtain New York State Department of Transportation (NYCDOT) certificates. During the 7:00 to 9:00 A.M. period, approximately 700 private commuter van trips are made into the Manhattan CBD, carrying about 15,000 persons per day.

## REFERENCES

Institute of Transportation Engineers, *Transportation and Traffic Engineering Handbook*, ISBN-0-13-930362-6, ITE, Washington, D.C., 1982.

NYC Department of Transportation, *Private Local Bus Improvements Study: Final Report*, New York, 1987.

Polytechnic Institute of Brooklyn, *Urban Public Transportation: Technology and Planning*, Department of Transportation, Planning and Engineering, New York, 1971.

## ADDITIONAL READING

Highway Research Board, *Bus Use of Highways: State of the Art*, National Cooperative Highway Research Program Report 143, HRB, Washington, D.C., 1973.

Institute of Transportation and Traffic Engineering, *Urban Mass Transit Planning*, University of California, Berkeley, 1967.

Metropolitan Planning Organization, *Transit Development Program: West Palm Beach Urban Study Area*, MPO, Palm Beach County, Fla., 1986.

NYC Board of Education, *Study of Pupil Transportation Services in New York City*, New York, 1978.

NYC Department of City Planning, *Express Bus Route Policy Study*, New York, 1986.

Pacific Institute for Public Policy Research, *Urban Transit: The Private Challenge to Public Transportation*, San Francisco, 1985.

Polytechnic Institute of New York, *Commuter Van Service Policy Study*, New York, 1985.

## QUESTIONS

1. Nationally, what percentage of all transit riders are carried by bus?

2. What are the various types of transit buses in terms of their size, seating capacity, design, and so on?

3. What are the advantages of bus transport over fixed-rail transport?

4. What are the four basic types of bus service available?

5. What are the different categories of bus service available?

6. What are the various service standards and categories typically used to evaluate bus transit?

7. What are the various forms of bus priority treatments in use today? Describe each briefly.

8. What criteria are used for the placement of bus stops?

9. What are the various components of bus operating costs?

10. What are the three major expense account categories that account for the majority of bus operating expenses? Describe each.

Mohonk

# 11

# PARKS AND RECREATION

## Adrienne G. Bresnan
NYC Department of General Services

## Joseph P. Bresnan
NYC Landmarks Preservation Commission

## HISTORICAL BACKGROUND

Most modern cities, towns, and even small villages have a park or open space set aside for the purpose of public gathering, beautification of the civic center, sports activities, picnicking, and other forms of recreation. Town squares, parks, playgrounds, designed landscapes, and natural lands, woods, beaches, and forests outside the town proper exist for these uses almost everywhere in the world, and have existed in some configuration virtually throughout history.

In ancient civilizations the element we now recognize as standard components of an urban park system were strongly in evidence; however, they may or may not have been available to everyone in past societies in the ways that are familiar to us today. The Roman forum and the Greek agora were central open squares used for public gatherings, speeches, announcements, and the like. Arenas, fields, theaters, and circuses for games competitions and other public entertainments were features of even relatively small cities. Gardens, baths, and spas were created at special locations, taking advantage of vistas, the sea, water, or extraordinary places. These locations necessitated varying degrees of design, engineering, architecture, and improvement, and as seen in this context, are early examples of "park infrastructure."

Old world influences were transported to the Americas by colonists from various nations who established their earliest and most rudimentary settlements using traditional planning layouts and schemes during the seventeenth century. As the colonies evolved into a nation through the eighteenth and nineteenth centuries, American public squares and commons required greater definition and outfitting for communal uses. For example, the very surface of the ground may have needed to be leveled and made firm for use of the local militia in drilling. A bandstand might have been erected to provide a stage set above the gathering audience for musical events. Shade trees may have been introduced to provide cover for strollers on sunny days. These were the beginnings of infrastructure, the physical improvement of places in response to the specific ways in which these places were to be used. By the 1850s the growth of cities and the impact of the industrial revolution raised serious concerns for the urban environment and the stresses it placed on the population, particularly in urban areas, where high density brought about conditions dangerous to public health, both physical and mental.

In a city like New York, development was following a formally established gridiron plan of streets and avenues relentlessly filling open space with buildings, leaving little, if anything, in the way of squares and parks. As this threat was recognized, social thinkers and planners of the time sought ways to modify growth patterns and set aside adequate space, dedicated and improved for public use. Thus was born the contemporary concept of the public park and the very essence of the term "to recreate" in association with the benefits of nature—

meaning the elements of landscape, affecting public health, well-being, and enjoyment.

With the design and construction of Central Park by Frederick Law Olmsted and Calvert Vaux in New York City during the 1850s and 1860s, the park movement in America was truly born. From its inception Central Park was a national prototype of such extraordinary scope and intent, based on fundamental principles of beauty and function, that it has yet to be surpassed (Olmsted, 1870). Central Park was the wellspring of a parks movement that would spread across America, and for the century that followed, landscape architects, urban planners, civil engineers, environmentalists, and a host of specialists would take their inspiration from the work of Olmsted and Vaux, adding their own concepts, solutions, and refinements in the creation of parks and park systems in cities and natural regions across the country.

## TYPES OF PARKS

The vast legacy of properties that have been reserved and constructed for park and recreation uses requires us to develop broad divisions and definitions to assist our understanding of the scope and meaning of parkland as infrastructure. One hierarchy distinguishes among national, regional, and urban parks; another separates natural resource areas from landscaped parks and special facilities.

National parks were started in the 1870s largely as a result of some imaginative photographers, such as William Henry Jackson and Carlton Watkins, who depicted the splendors of the west and brought photographs back to Congress with the message that this priceless heritage had to be preserved and made available to the public. National parks such as Yosemite and Yellowstone were developed on land remote from urbanized areas; but over the years our extreme mobility has made them so easily accessible that their survival as a once-annual refuge from the pressures of civilization has been placed in doubt. Their functions include camping and hiking, recreation, and the study of flora and fauna.

Regional parks were created in areas much closer to cities. Walking trails, picnic areas, and beaches are available to the city dweller on a weekly basis. Foresight by state governments, and in some cases the generosity of public-spirited patrons, have placed this land into the public trust.

It is the urban parks, however, which provide the amenities that make life in cities possible. Golden Gate Park in San Francisco, the lakefront in Chicago, and Central Park in New York not only provide vast reservoirs of light and air, but offer an escape from the densely settled neighborhoods around them. Although large urban parks make it possible to provide a wide variety of functions, even a single block of open space can serve as a gathering place, similar to the village green of yesteryear. To watch central city office workers spill into all available open spaces during lunch hour on a beautiful day is a clear indication of the inherent need.

All types of parks contain natural resource areas; out of the 26,000 acres that make up New York City's parkland, 25% are in their natural state, and they are the most difficult to maintain. Human evolution and science have affected the natural environment greatly, leaving little of the world untouched, or indeed, "forever wild." In the largest sense mankind is faced with managing the world environment in response to the results of human activity. The more access people have to natural resource areas, the greater the need for protection and control. Facilities must be designed for human enjoyment, safety, and comfort, as well as for the preservation and conservation of the resources themselves. The living elements of such areas, the plant and animal populations, and the complex settings on which they depend, need the greatest care and management. This aspect of the environment should be understood to be a category of park infrastructure which we must learn to comprehend, to monitor regularly, and which we must maintain and restore to ensure its conservation.

Naturalistic, constructed landscape parks, on the other hand, are more readily recognized to be a form of infrastructure because their history can be traced. The grading and shaping of the land, establishment of water features, plantations, drives, and walks, and the networks of utilities providing irrigation, drainage, power, and light are all discernible as works of civil engineering and landscape architecture. Such recreational environments have their own set of strengths and weaknesses with respect to our ability to discern their performance under the wear and tear of heavy use throughout the seasons. As we have seen, the idea that "nature will look out for itself" no longer holds true for natural areas, much less for constructed parks, which are generally built in urban areas. The engineered features of a naturalistic park are of course susceptible to the problems of other types of systemic infrastructure, but deterioration and failure of a park are sometimes subtle and imperceptible until critical (Stiran, 1984). Those entrusted with the design, construction, maintenance, and conservation of parklands need to educate the public about the special qualities and frailties of parks in order to establish and keep them properly for the enjoyment of everyone.

As we move from the less structured to the more structured park setting, the specialized fields, courts, tracks and courses, ranges, pools, and other types of sports facilities form an important and complex set of components of a park system. Although many of the facilities have specific dimensions and physical layouts or equipment for a particular sport, collectively they form a category of landscape and contribute to the open space in unique ways. Expanses of grass and turf in the cases of golf courses or ballfields, specially shaped and surfaced areas such as tennis courts and running tracks, or water sports facilities such as racing and diving pools contribute structure and function to recreational parks

(Wurman et al., 1972). Recreational sports facilities require the building of structures to accommodate the performance of the sport and the seating of spectators, including buildings, stadiums, and bleachers, along with all the ancillary services, such as power and light, drinking water, toilets, refreshment vending operations, and other installations meeting human needs in a comfortable and enjoyable way.

In addition to sports attractions, a park might contain a zoological collection, aquarium, botanical garden or arboretum, or theater or other performing facility. Such highlights of a park system can vary widely in scale and significance, from world-class institutions involving advanced academic work and research and entertainment endeavors, to simple children's farms or ornamental gardens and plantings and puppet shows. Obviously, this variety of improvements can involve great investment and expense, as well as potential revenue, and brings up the broad and controversial subject of the appropriateness of nonpark activities such as "theme parks" and "amusement parks" within the context of the more traditional and sacred open-space conservation concept and policy.

Competition for open space is keen and there is a wide range of perceptions and sensitivity on the subject, which the wary practitioner must learn about, because it goes to the heart of what parks mean to people. Most jurisdictions manage to strike a balance and try to respond to diverse constituencies on the national, state, and local levels. The great diversity of sports, recreational, and institutional settings frequently found within larger natural areas and landscape parks gives some indication of the range of skills and specialties required to design and build such facilities, and then to operate and maintain them. Numerous professional disciplines and many specialists, mechanical tradespersons, and skilled maintenance workers are required to create and care for park and open-space infrastructure.

## CRITERIA

Criteria have been developed for the design of park facilities for newly developing areas, or to assess the adequacy of existing open spaces. A general statement by the National Recreation and Parks Association recommends that a minimum of 25% of new towns, planned unit developments, and large subdivisions be devoted to park and recreation lands and open space. More specific criteria are summarized in Table 11-1. In addition to size criteria it is important to know how far from a residence such recreational facilities should be located. Recommended distances are listed in Table 11-2. These numbers represent a useful starting point for planning. Obviously, needs as well as the capability of meeting them will vary widely.

**TABLE 11-1  PARK AND RECREATION CRITERIA**

| Type of Facility | Population Standard | Site-Size Standard |
|---|---|---|
| Major natural park | 1/40,000 | 100 acres/park |
| Public golf course | 1 hole/3000 | 100 acres/18 holes |
| County fairground | 1/county seat | Varies |
| Public stadium | 1/100,000 | Varies |
| Botanical garden | 1/metro area | Varies |
| Zoo | 1/metro area | Varies |
| Playground | 1 acre/800 people | 3–6 acres |
| Local park | 1 acre/1000 people | 2 or more acres |
| Recreation center | 1 acre/800 people | 15–20 acres |
| Playing field | 1 acre/800 people | 10–30 acres |

Source: F. S. Chapin; Urban Land Use Planning, Univ. of Illinois Press 1963

**TABLE 11-2  TIME–DISTANCE STANDARDS FOR URBAN AREA OF 100,000 POPULATION**

| | |
|---|---|
| Playgrounds and local parks | $\frac{1}{2}$ mile |
| Playing fields and recreation centers | 1 mile or 20 minutes |
| Public park or reservation | 30 to 60 minutes |

Source: F. S. Chapin; Urban Land Use Planning, Univ. of Illinois Press 1963

## SUPPORTING INFRASTRUCTURE

Parks form a part of the urban infrastructure in and of themselves; however, they must be supported by many of the elements described in this book. This includes water supply, drainage facilities, utilities, and means of circulation. To provide essential human services in natural-area parks or special facilities, a wide range of buildings and structures has been found to be necessary, and a number of special-purpose and multipurpose concepts have been developed.

The important quality desired in park architecture of any kind is its unobtrusiveness and efficiency. that is not to say that buildings in parks should not be distinctive, but rather that they should serve their intended functional purpose and further enhance the essence of the park experience either by receding into the landscape, allowing it to predominate, or by contributing to the experience through the appropriateness and spirit of the design.

Buildings in parks can be celebratory of their function, such as a performance pavilion, gazebo, grandstand, or greenhouse, splendid examples of which can be found in historical as well as contemporary design. Buildings in parks are generally designed for specific purposes of short duration, such as recreation programs, food and refreshment vending, locker and shower facilities, performances, and toilet and maintenance services. Each of these functions is desirable and necessary and therefore requires clarity of use-related design and signage, "barrier-free" access, and overall efficiency of operation and energy cost.

Another major category of structure commonly found in parks are bridges and arches. Bridges serve to separate various forms of circulation within parks or provide a means of overcoming one or another kind of obstacle to circulation, such as streams or a highway. Often a bridge can offer a special vantage point to overlook a scenic view. Although functional and utilitarian structures, bridges and arches can be designed as unique and distinctive architecture serving as ornamental features in the park setting through the use of form, material, and detailing with special character.

In older parks and parkways, a rustic style was frequently used for bridges and arches wherein natural boulders or coarse stone were utilized, evocative of a rural or alpine setting. Eclectic forms of architecture were used in Victorian-era park environments. In Central Park, for example, and later in Prospect Park, Calvert Vaux and his design team embellished bridges and arches in decorative styles representing motifs drawn from many parts of the world, including Europe, Asia, and Africa. Within the park landscape, bridges serve secondarily as shelters from the sun or a rain shower. Arches provide transitions in park design which add variety, surprise, and mystery to the experience of the designed environment (Laurie, 1977).

Parks and park systems contain roads, paths, drives, and arteries accommodating vehicles and pedestrians in functional as well as recreational ways. Scenic parkways have been designed to take advantage of scenic natural areas offering motorists opportunities for "pleasure driving." Similarly, landscaped parkways, drives, and avenues, in and of themselves, afford motorists and pedestrians special experiences not found along expressways or ordinary streets where minimal or no vegetation at all is provided to screen unattractive views.

Equestrian roads and paths are sometimes introduced into parks and along arterial roads. Concern arises with the intermingling of different forms of circulation, where roads and paths need to cross or interrelate. Bridges and arches, fencing, signage, signalization, and other forms of separation and control are required which necessitate careful design to be both functional and esthetically pleasing.

In New York City, Olmsted and Vaux first introduced the important concept of linking major parklands by tree-lined boulevards called "parkways," the earliest being Ocean and Eastern Parkways, creating the framework of the Brooklyn park system (Fein, 1968). Such sweeping city planning concepts were to have a great influence on later systems of parkland as they were acquired and developed. The twentieth century would see replacement of the carriage and equestrian traffic with the introduction of the higher-speed automobiles. Modern parkways were no longer grand boulevards of the urban street pattern, instead becoming landscaped scenic roads offering pleasure driving through richly planted and often spectacular scenic points of the city.

Although the parkways have been progressively adapted to serve faster traffic flow and safety needs, and have been rebuilt to standards other than those of their original esthetic and scenic intent, their importance in linking and interconnecting the parks of a city together cannot be underestimated. Parkways and green corridors provide more pleasant driving for motorists, buffer residential and recreational areas, and allow for the creation of trails for walking, cycling, and jogging, which have become major recreational forms for large numbers of park users across America.

As infrastructure, park roads and walks present all the same concerns encountered with arterial systems elsewhere, such as surface maintenance, drainage, safety, and functional capacity. However, the special concern for these installations with regard to parkland is their subordination to the greater intent of esthetic and recreational experience—to have beauty and give pleasure—not the efficient movement of vehicles and traffic.

The more successful a park design, the less the user is aware of the several systems that have been integrated into the plan for the user's convenience, safety, and enjoyment. Basic to any constructed landscape or controlled environment is some form of drainage system. Access paths in natural areas, where minimal intervention is uppermost in the minds of the designers, must involve the redirection of water runoff and careful selection of materials, gradients, and contours. Bodies of water or naturalistic water features in designed parks often involve control of both water supply and overflow, in addition to proper handling of groundwater and erosion in areas affecting the feature. Adequate ditches, swales, drain piping, catch basins, and sewers are required to be installed and properly maintained to sustain a landscape park, recreational sports field, courts, and so on.

Parks and recreational areas, especially in urban settings, require some amount of lighting sufficient for passive or active use levels. Roads and paths might be illuminated, athletic fields and sports facilities may be equipped with night lighting; buildings, bridges, and other facilities may need power for lighting, equipment and heating. These electric utility installations, like the plumbing for water and gas, or the wiring for telephone communications and security systems, need skillful installation and upkeep in order to provide service without detracting from the park design. Removal systems for wastewater, garbage, and debris from park maintenance work also necessitate thoroughly designed solutions to prevent these essential services from presenting health or safety hazards to park users, or from becoming detractive to the natural environment or park facility in a visual or other noticeable way.

By extension, park utilities will include the vehicles used for routine grounds keeping and repair, and for the gathering and removal of trash and horticultural debris, as well as the garages, shops, personnel, and storage facilities that form operations and maintenance services. These services are critical to the cyclical maintenance of a park and are frequently housed in the park area to be on hand throughout the seasons of the year.

## PLANNING METHODOLOGIES

In planning for the rehabilitation of an existing park or the development of a new one, the methodologies are similar. The first step consists of obtaining a base map showing all the parcels in the vicinity of the site being studied. This is followed by a site survey to assess the suitability of the area to support park development or to determine the condition of an existing facility. The next step in launching a construction or rehabilitation program of significant proportion, involving major capital assets, is to state or restate, as the case may be, the principles and policies governing the program. This is in line with conventional planning practice of setting goals and objectives.

Based on a knowledge of what recreation facilities exist in the study area, their carrying capacity is calculated on the basis of accepted standards. This is then compared to the estimated requirements and forms the basis of the proposed program. Combined with the stated policy, the program then becomes the recommended master plan.

The authors have had the opportunity to initiate massive reinvestment projects for the New York City Park System as well as for a number of its important regional parks. They have drawn on their extensive experience to present excerpts from a number of documents as they relate to their long-time involvement. Exhibit 11-1 represents an inventory of existing parks and recreation infrastructure for the city of New York. Exhibit 11-2 is an example of how policy-planning issues are dealt with. "The Green Plan for Parks—1985" is cited as a proposal developed to define and give context to an agency-wide "greening" program expected to take priority within the overall capital commitment plan of the agency for an extended number of years. It attempts to restate the values and objectives of the Department of Parks and Recreation concerning the totality of its physical plant or its aggregate infrastructure, including the assertion that there is such a thing as a "living or green infrastructure" which forms the critical and unique mandate of an agency entrusted with parkland (Fox et al., 1985).

Exhibit 11-3, the table of contents for the Central Park Master Study of 1973, represents a methodology that was the most comprehensive and farsighted planning tool yet devised to comprehend the condition of such a unique historical artifact and living environment. It established the park's needs as a matter of municipal priority and budget commitment for the next generation. Exhibit 11-4 gives a sense of the analysis required to assess the deterioration of the park and to determine the physical needs of a complex park infrastructure that had been serving the public for over 125 years.

Since funds for rehabilitation are rarely adequate for the tasks at hand, private park support organizations have raised funds to support repair and maintenance projects; but more important, these citizen groups have actually undertaken the work of planting and grounds maintenance themselves. Some of the statements and proposals outlined in these exhibits seem simplistic and matter of fact. However, before they were written, these concepts were by no means the criteria used for decision making or reason for action by the city. Only by reexamining fixed assets and determining value is policy given definition, context, and commitment (Cranz, 1982).

## ACQUISITION AND COSTS

Parks are created or acquired in many ways. National parks were initially staked out by the federal government; recent acquisitions have been through purchase out of federally budgeted funds. The Nature Conservancy, a national organization dedicated to the preservation of open land, has been the beneficiary of wills made by large land owners. The conservancy in turn gives the land to the federal government in exchange for a guarantee of keeping it forever wild.

**TABLE 11-3   SELECTED PARK AND RECREATION DEVELOPMENT COSTS**

| | |
|---|---|
| Large regional park | $200,000 to $300,000/acre |
| Small urban park | $40,000 to $60,000/acre |
| City playground | $600,000/acre |
| Running track, 400-meter standard | $700,000 each |
| Playing field | $250,000 each |
| Tennis court | $65,000 each |
| Recreation center | |
|    Small | $300/sq ft |
|    Large | $150/sq ft |
| Maintenance garage | $275/sq ft |
| Street trees, 3- to 4-in. caliper | $400 to $500 each |

Funds for the acquisition of parkland are part of general budgets at all levels of jurisdiction. Acquisition can be by condemnation or by negotiated purchase, with compensation determined by fair market value. On the local level it has become an established practice to require developers of more than three or four homes to dedicate a percentage of their land for recreational purposes as part of the subdivision regulations. It is generally permitted to avoid such a dedication by making cash contributions to the local recreation fund. In the suburbs this has worked well in that both the land and the funds to develop it for recreational purposes have been made available. Representative costs for park and recreation development are listed in Table 11-3.

## REFERENCES

Cranz, Galen, *The Politics of Park Design*, MIT Press, Cambridge, Mass., 1982.

Fein, Albert, *Landscape into Cityscape: F. L. Olmsted Plan for a Greater New York*, Cornell University Press, Ithaca, N.Y., 1968.

Fox, Tom, Ian Koeppel, and Susan Kellan, *Struggle for Space: The Greening of New York City, 1970–1980*, Neighborhood Open Space Coalition, New York, 1985.

Laurie, Michael, *An Introduction to Landscape Architecture*, University of California, Berkeley, Elsevier, New York, 1977.

Olmsted, Frederick Law, *Public Parks and the Enlargement of Towns*, Riverside Press, Cambridge, Mass., 1870.

Stiran, Anne Whiston, *The Granite Garden: Urban Nature and Human Design*, Harper & Row, New York, 1984.

Wurman, Richard Saul, et al., *The Nature of Recreation: A Handbook in Honor of F. L. Olmsted*, MIT Press, Cambridge, Mass., 1972.

## ADDITIONAL READING

Anold, David, ed., *Practice of Local Government Planning*, Municipal Management Series, International City Managers Association, Washington, D.C., 1970.

Boston Foundation/Carol R. Goldberg Seminar, *The Greening of Boston: An Action Agenda*, Boston, 1988.

Fein, Albert, *F. L. Olmsted and the American Environmental Tradition*, Braziller, New York, 1972.

McHarg, Ian, *Design with Nature*, Natural History Press, New York, 1969.

Newton, Norman, *Design on the Land: The Development of Land Scape Architecture*, Harvard University Press, Cambridge, Mass., 1978.

State of New York, Parks Recreation and Historic Preservation, *People, Resources, Recreation 1983: NYS Comprehensive Recreation Plan*, Albany, N.Y., 1983.

## QUESTIONS

1. Who is served by the establishment of park and recreation facilities?

2. Discuss the kinds of park and recreation attractions found in a typical community.

3. How would you determine the scope and scale of a park system for a municipality of 50,000 population?

4. What concern would you have in the protection of natural parklands encompassing a forest and wildlife region?

5. What importance do parks have in relation to other forms of infrastructure and public services in urban areas?

6. How do trees and plants, the living green infrastructure, make a difference in cities?

7. Can you define the term *design with nature?*

8. As a park and recreation administrator, how can the community you serve help you do your job?

9. Could you provide your own definition for the concept "parks are the lungs of the city"?

10. How can parks and open space be used as teaching tools to educate park users of all age groups?

11. What arguments would you use to obtain the funding priority for the acquisition, construction, and maintenance of your parks and recreation program?

Exhibit 11-1

**Exhibit 11-1** Department of Parks and Recreation infrastructure chart. (Statistics from NYC Department of Parks and Recreation.)

Department of Parks & Recreation Infrastructure Chart

Area

| | |
|---|---|
| 24,610 acres | 12% of city's 194,708 acres |
| 572 parks | |
| 900 playgrounds | 1472 total of parks and playgrounds |
| 300 miscellaneous sites | Malls, squares, and triangles |
| 890 playing fields | Baseball, softball, football, soccer |
| 526 tennis courts | 3,210,000 sq ft or 73.69 acres of playing surface based on standard 50 × 120 ft court |
| 19 running tracks | Olympic, borough, and neighborhood facilities with track and field events |

Recreation Facilities

| | |
|---|---|
| 7 ice rinks | 2 indoor: Abe Stark and Flushing |
| | 5 outdoor: Wollman (Manhattan and Brooklyn), Mullaly (Bronx), Lasker (Mannhattan), and Clove Lakes (Staten Island) |
| 13 olympic pools | |
| 24 intermediate pools | Indoor and outdoor pools and support facilities |
| 35 minipools | Outdoor neighborhood pools |
| 34 recreation centers | Programs for youth, preschool, senior citizens, handicapped, etc. |
| 13 golf courses | |
| 6 beaches | 14.9 miles, some with boardwalks |
| 3 city zoos | Central, Prospect, Flushing Meadows |

| Street and Park Trees | *Boro* | *Street* | *Park* | *Total* |
|---|---|---|---|---|
| 2,600,000 trees | Bronx | 53,000 | 366,000 | 419,000 |
| | Brooklyn | 127,000 | 120,000 | 247,000 |
| | Manhattan | 34,000 | 190,000 | 224,000 |
| | Queens | 305,000 | 1,120,000 | 1,425,000 |
| | Staten Island | 81,000 | 204,000 | 285,000 |
| | | 600,000 | 2,000,000 | 2,600,000 |

Operations

7 administration headquarters and 16 garages citywide
1500 vehicles and equipment, including mobile recreation units

**Exhibit 11-2** Green Plan for Parks, 1985.

## I. Introduction
## Summary Policy

The Department of Parks and Recreation has a primary mandate unique among the agencies of the New York City government. It is charged with establishment and maintenance of public parks throughout the city, a responsibility clearly explained in the Charter, which refers to parks in terms of beauty and utility, and requires that these assets be conserved and improved. Parkland is protected, set aside in trust for the enhancement of the city and enjoyment and health of its residents.

Natural and designed open space forms the essence of the park infrastructure. Presently, nearly 26,000 acres of the City of New York is designated parkland, constituting approximately 20% of the five boroughs. This acreage encompasses historic city squares dating from the earliest settlement of the city, spaces that mark its growth, and great designed parks which symbolize the leadership and imagination of its civic achievement. Within the system may be found recreation, diversion and enjoyment to satisfy the broadest possible usership. It is a complex amalgam of properties bound together by a common purpose—to provide an escape from the rigors of urban life, a moment to reflect in a place apart from the noise and hustle of city life. While the city offers many non-park recreational opportunities of every description, the millions of people come to our parks, beaches and open space facilities to relax, play ball, sit, picnic, etc. What is it about parks that makes them so special?

*Parks Are Green.* They are soft places that are filled with living plants and animals, the only vestiges of nature preserved in the midst of a vast structured human environment. It is this 'pre-eminence' of what is green-trees, shrubs and grass in sweeping landscapes with dramatic rock and water, or in a local playground, triangle or mall-that is and must be the central mission of the Department of Parks.

The City of New York has had an impressive record in the creation of parks. Strong constituencies in the public and private sectors have played important roles in the past. But never before has there been so broad a base of support, so keenly interested and knowledgeable a parks group and so many demands on the city to improve services and protect our precious parkland. These factors combine to make this a special moment in time to make great strides in park improvement and care.

•• The enormous value of our park system needs to be readdressed as an encompassing Capital asset. The parks themselves, including landscape, forests, trees, and horticultural stock, fields and meadows, water bodies, facilities and utilities, etc., are major categories of permanent infrastructure.

•• The quality of scenic open space areas must be reinforced and preserved for the intrinsic beauty, environmental and psychological function they provide as a relief and alternative to urban density. The essence of this quality is the access to open space, vistas and botanical nature readily available to the neighborhoods of the city.

•• The scenic and green linkage value of the historic and modern parkways needs to be reaffirmed and restored. Enhancement of Parkways, major avenues, boulevards and

Exhibit 11-2

principal streets with cohesive street tree plantings would have a significant effect on the urban design and environmental quality of the city.

•• Neighborhood Street and Park Tree programs, community gardens, ornamental horticulture at important civic sites all enrich and deepen the liveability of the places where we live and work through the use of plants.

•• To the greatest practicable extent, pavements and structure in parks must be minimized, while turf areas, plantations and trees are maximized throughout the park system.

•• The Department of Parks and Recreation needs to secure the alliance, cooperation and support of the Community Boards, block associations, city-wide and community-based parks constituency groups, botanical, horticultural and gardening organizations, environmental groups for the purpose of engendering a network of 'greening' forces dedicated to preserving our great parks system, enlarging and enriching the urban forest, embellishing public squares and civic areas with ornamental plantings, etc. We must also strengthen cooperative efforts with the National Park Service and New York State Parks and Recreation pertaining to their parks jurisdiction within New York City.

•• We need to extend and acquire additional open space accessible to the *waterfront* in the form of natural shorelines, promenades and esplanades, recreational piers, observation points and general public access to the edges of our city. New York City, after all, is an Atlantic sea coast port consisting of a series of islands and shorefront lands.

## II. The "Green Plan"

The Department of Parks has evolved a "Green Plan" for the restoration of the entire five borough parks system to be implemented through the Capital Budget process over successive fiscal years. The program is determined through the collaborative efforts of the agency with the 59 Community Boards of the city, aided by a broad and vocal city-wide parks constituency and political leadership.

The Capital Budget of the Department of Parks and Recreation offers opportunities to acquire, stabilize, improve and conserve parklands, natural resources and recreation facilities. In re-focusing capital priorities, the agency has placed increasing emphasis on the fundamental landscape infrastructure of the parks, consisting of the tree stock and plantation, forest, hills, meadows and fields, drainage systems and pedestrian means of access and interpretation. Within this emphasis on the naturalistic is the general directive to remove obsolete and excessive pavements and structures where they are found to be inappropriate and unnecessary maintenance burdens. Preservation of mature healthy trees and plantation is paramount, along with stabilization and rehabilitation of the landscape, erosion control, and new permanent trees, plantations and lawn areas. Rehabilitation of existing, sound community recreation facilities will be undertaken as required to maintain organized programs and services.

Stabilization and reclamation of water bodies, shorelines, and wetlands are essential aspects of restoring major parks which contain such features. Dredging, protection against flooding and pollution, illegal dumping and abuse, are all important items of work in preserving natural areas and restoring them to optimal beauty and function.

The "Green Plan" is a series of interrelated and comprehensive Capital Projects that have been funded in the current and future Capital Budget of the City of New York. The projects are listed in the following categories for each borough as major components of the commitment plan, including projects in planning and those actually under design.

- Landscape parks— "flagship parks"
- Waterfront parks
- Small parks and playgrounds
- Street trees and parkways
- Parkland acquisition
- Major items of park maintenance equipment
- Other park jurisdictions

- In each borough significant regional *landscape parks* are focal points of the capital program. These 'flagship parks' include some of the greatest historic landscapes and largest inland recreational parks in the system. These parks often contain major water bodies of scenic and environmental importance, and established landscape features in deteriorated or weakened condition. These primary landscape parks must be restored to continue serving the largest population of park users.

- The borough programs contain strong emphasis on *waterfront parks* including beaches, waterfront facilities, designated wetlands, esplanades and undeveloped property (including landfills) of great potential.

- *Small parks and playgrounds* having the greatest accessibility to neighborhoods make up a large share of the city-wide program to provide functional facilities where they are most needed.

- *Street Trees and Parkways* is a category of projects including the greening and rehabilitation of parkways, boulevards, avenues and streets. Such projects include connective landscapes and linkages such as Pelham Parkway and Ocean Parkway, thoroughfares such as Woodhaven Boulevard, and street tree replacements and additions virtually anywhere in the city. The Capital Program may also include contracts for removal of dead and dying street trees, stump removal and heavy pruning.

- *Parkland acquisition* is another form of Capital Project, where outright purchase is required. The department has negotiated a number of important acquisitions, particularly for the Greenbelt in Staten Island, obtaining important additions before development pressures and land values made it impossible. Other potential acquisition sites are in the listing.

- The Capital Budget also covers purchases by the city of *major items of equipment* having a value over $15,000 and a life expectancy of at least 5 years. This category includes fixed equipment such as boilers, pumps and mechanical equipment, and large vehicles such as high rangers for tree work, etc.

- Each borough listing includes notation of parkland under the *jurisdiction of other agencies* such as the National Park Service and New York State Parks and Recreation, which provide significant services to New Yorkers augmenting and extending the city's own resources. In most cases substantial improvement programs are underway at these sites as well.

Exhibit 11-2

### III. The Green Future
### (Expense Budget)

*Natural areas and living things go through a continuous cycle of renewal.* In the urban environment the enormous stresses upon survival can prevent or impede the renewal cycle resulting in poor development or early demise of plant materials and failure to achieve growth and the intended scenic richness of a landscape design. Street Trees are perhaps the most threatened of all urban plant installations especially in the densest situations. Special selection, maintenance and replacement when necessary are required for the success of trees, shrubs, ground covers and turf under difficult circumstances.

• An urban park or plantation is part of the infrastructure and capital establishment of the city requiring cyclical maintenance and resource management if the value of such physical assets are to be retained. Appropriate allocations in the Expense Budget must be made corresponding to renewed capital improvement of the landscape infrastructure, i.e. more skilled tree and garden care personnel, equipment, materials and a plan to use these resources correctly.

• Urban beautification programs are entirely feasible only when the maintenance implications are understood and addressed. Seasonal requirements must be met and critical periods of the annual calendar must be committed to predetermined needs of the living inventory.

• Training programs are required to teach the skills and familiarity required to care for plantations of all kinds, natural and ornamental, and of the support systems necessary in the urban situation—water, nutrients, pest controls, aeration, pruning, etc.

• Public education and orientation programs are needed to sensitize individuals to the qualities of the urban environment, the needs of growing trees and plants, their virtues and limitations. Often simply intelligence and assistance can be of enormous help.

• The cooperation of other agencies is mandatory if the long term establishment and maintenance of landscapes and trees is to be seriously pursued. The Department of Transportation, for example, must become a willing ally of the Parks Department if street trees, parkways and roads through parks are going to be part of our future. A Parkway Commission is recommended to achieve this goal.

• The function of trees, plants, grass and landscape in our urban lives needs to be better understood and emphasized. A living environment is more than attractive, it has real psychological effect on most people and a positive impact on their daily lives.

• The Department should reclaim its rightful role of leadership in the City for all matters of *"greening."* A century ago botanical excellence was to be found in the greenhouses, conservatories and gardens of our park system. The great institutions have since taken the lead while basic and rudimentary horticultural skills grow ever scarcer in the Parks Department.

• The Department should seek the help of qualified institutions and societies and groups to re-establish skills and knowledge within the department, using the vast resources and outdoor laboratory of the parks to learn with. The vestiges of a great horticultural past are still to be found and should be restored, greenhouse operations, nurseries, gardens, planters, flower beds, orchards, and splendid landscapes all await revival.

• Restorations and installations by Capital means can only go part of the way to restoring our parks. A strong, skilled and adequate staff is necessary for total success.

• Volunteerism can be of major assistance to the Department if well used and managed. However, the same rule applies to volunteer efforts as it does to capital investments-follow up committed maintenance is essential if the improvements are to be permanent.

Exhibit 11-3

**Exhibit 11-3** Central Park Master Study: comprehensive listing of study areas from the architects' office of the Central Park Task Force.

Outline for Central Park Master Study

A. Interagency Operation and Responsibilities
  1. How do agencies operate within Central Park?
  2. Is there coordination with PRCA Central Park Maintenance and Operation staff?
  3. Does PRCA have consolidated information concerning all the City agency services?
  4. Are there master surveys of telephone lines, steam lines, etc.? And does PRCA have access to this information?
B. Mechanical and Physical Features of Central Park, excluding Buildings
  1. Arterial traffic system
    a. Description of system
      (1) Drives
      (2) Transverse roads
      (3) Bridle path
      (4) Pedestrian paths
    b. Automotive, pedestrian, and equestrian
    c. Entrances and exits and four corner conditions
    d. Precise extent and condition
    e. Necessity and origin
    f. Parking facilities
  2. Bodies of water
    a. Description of bodies of water
      (1) Edge
      (2) Bottom
      (3) Geography
      (4) Closure treatment
      (5) Use pattern
    b. Plumbing and control
    c. Seasonal operation
    d. Water shed
      (1) Erosion control
      (2) Drainage
    e. Water characteristics
      (1) Volume
      (2) Depth
      (3) Quality
    f. Reservoir as atypical condition
      (1) Justification
      (2) Potential public utilization
      (3) Continuation of control apparatus
    g. Precise extent and general condition
    h. Necessity and origin

3. Earth and rock contour
   a. 1936 topographical survey as guideline and, additionally, any historical data available
   b. Existing cross sections
   c. Natural rock out-croppings
   d. Update topographs with contemporary land survey
   e. Borings, test pits, soundings, soil investigation, excavations
4. Utility systems
   a. Electrical: metered and unmetered
      (1) Street lamps
      (2) Park lamps
      (3) Building services
      (4) Spot, feature, and emergency lighting
      (5) Safety lights
      (6) Trunk lines: subsurface conditions
      (7) Necessity and origin
      (8) Precise extent and condition
   b. Plumbing: potable water supply for:
      (1) Buildings
      (2) Drinking fountains
      (3) Display fountains
5. Sewage: sanitary drainage
   a. Buildings
   b. Fountains
   c. Pools, rinks, and zoos
   d. Location of any water systems circulating under park to and from sources unrelated to the park
6. Drainage: water run off for surface water shed
   a. Catch basins, swales, culverts, etc.
   b. Investigate all water collecting devices
7. Telephone services
   a. Locate building service, wiring layout and source of service
   b. Police call boxes
   c. Fire call boxes
   d. Investigate any major trunk lines for the city located in the park
8. Gas supply
   a. Extent of use in park
9. Steam supply
   a. Arsenal and zoo complex
   b. Other usage?
C. Special Occupants in Park
   1. Weather stations
      a. Arsenal
      b. Clarify extent at Belvedere

Exhibit 11-3

2. Transformer enclosure near Belvedere Terrace
   a. Investigate any other devices in the park
3. Fire communications at 79th Street transverse road
   a. Definition of operation
   b. Status of proposed new underground structure to be constructed at eastern end of Belvedere Lake
4. Police Station, 22nd Precinct, Park Police
   a. Definition of operation and extent of services
5. Subway
   a. Investigate existing subway lines under Central Park South and Central Park West
   b. Investigate new subway construction
   c. Subway structures
      (1) Ventilator and emergency exits
      (2) Subway entrances and exits
      (3) Central Park South wall ventilators
      (4) Subway gratings, exhaust devices in park sidewalk, walls, free standing, etc.

D. Operations of Park
   1. Maintenance, operation, and concession services
      a. Deliveries and removals
      b. Materials and supplies
      c. Storage
      d. Garbage and waste
      e. Oil and gasoline

E. Buildings
   1. Survey and categorize all buildings and structures within master park plan
      a. Size
      b. Style
      c. utilization
      d. Site location
   2. Determine present conditions, deficiencies and assets
      a. Weathertight evaluation
      b. Vandaltight security
      c. Utilities, etc.
   3. Aesthetic status
      a. Compatibility
      b. Architectural merit
      c. Historic significance
   4. Rehabilitation
      a. Phase I: determine ongoing use; comparative cost analysis
      b. Phase II: restoration, redesign, or demolition
      c. Phase III: staged programming of rehabilitation and redeveloped utilization of buildings, i.e., 86th Street shops conversion to riding academy and stables

d. Phase IV: assignment of physical restoration of viable structures
   (1) Immediate repair capability by PRCA personnel
   (2) Concessionaire capacity for general improvement of facilities
   (3) Restoration by contract procedure
   (4) In-house PRCA design capability outside consultant design input
F. Environmental Park Spaces and Furniture
   1. Determine the definition of spaces within Central Park
      a. Size, capacity, detail, scale, use, density, significance in the total scheme
      b. Master Plan Land Use Overlay to subdivide the topography (solely for the purpose of clear definition) into identifiable spaces. For example:
         (1) The formal promenade, Mall, and terrace
         (2) Four corner entrance circles to Central park: Grand Army Plaza, Columbus Circle, Fredrick Douglas Circle, Frawley Circle
         (3) Water elements in park
            (a) Pond
            (b) Lock
            (c) Gill
            (d) Belvedere Lake
            (e) Conservatory water
            (f) Reservoir (new public use pattern)
            (g) Harlem Meer
            (h) The Pool
            (i) Water shed landscape, land use, improvement of water resources
   2. Illustrative proposals as the outcome of simultaneous overviewing of the spaces with the most in-depth survey of specific physical contents of the park and use determination
   3. Coordination of new land uses with rehabilitated structure services, maintenance to insure future upgrade
G. Park Furniture
   1. Survey the park needs
   2. Survey present and local design
   3. Determine prototypical redesigns to function in park: benches, water fountain, fences, signs, utilities, paving, light standards

**Exhibit 11-4**

**Exhibit 11-4** Landscape and civil engineering rehabilitation. (From NYC Department of Parks and Recreation.)

Central Park is essentially a manmade picturesque landscape, the physical features of which have been rearranged and designed for specific spatial and atmospheric effects. By virtue of this physical conception, the Park was imbued with a romantic spirit, matched in quality and resonance only by its sequel in the Borough of Brooklyn, Prospect Park.

As the *first* Park of its kind in the United States, and as the supreme example of a uniquely American interpretation and distillation of the precepts of Romanticism expressed through nature and the manipulation of nature for the ideal, Central Park is a work of Art. This realization is omnipresent in the minds of those who shall attempt to preserve, restore and cherish it.

This proposal commences on the premise, evolved through months of extensive field investigation and inventory and research, that Central Park is in severely damaged condition not only in relation to its aesthetic heritage, but also in the fundamental reality of 840 acres of naturalistic terrain in the midst of the City of New York. The multifaceted and complex drainage system serving entirely diverse acreage is not functioning properly. Massive erosion poses multiple threats to the survival of horticulture, utilities, paths, bridle paths and roads in Central Park. The efforts of our maintenance operation are frustrated by constant emergency work which disrupts and protracts vital routine improvement, to the point where backlog becomes the *next* emergency. Increased manpower is one desperately needed measure which PRCA is strongly urging, but manpower alone cannot correct years of clogging, deterioration of material, loss of soil, mammoth plant management requirements and the obsolescence of facilities and mechanical service.

The first component of each rehabilitation Zone is the design and construction of all necessary drainage correction, land stabilization, and retention, regrading and earth movement restoration, and the conditioning of water bodies, banks and related landscape elements. Integrated with the civil engineering rehabilitation shall be the horticultural improvement, deletion and additions as necessary. All of the work will be designed and coordinated by professional Landscape Architects and a PRCA Special Central Park Office, all of whom shall be responsible for the aesthetic and practical decisions pertaining to the results of the entire Master Rehabilitation.

Within the enormous framework of Landscape restoration, closely coordinated plans shall be developed for the work required to rehabilitate the architectural structures in each respective Zone. Construction will be scheduled so as to complete the major building and bridge projects prior to the landscape and horticultural redevelopments. In the case of bridges in particular, the coordination will have the greatest priority to minimize the duplication or repetition of work.

Certain Zones bear considerable relationship to adjacent Zones. This is an obvious observation in perceiving the entire landscape, however for purposes of technical clarity the topographical factors defined and described Zones-with regard to contour, surface and subsurface drainage and arterial connections. Engineering and architectural decisions will resolve the relationships of staging the projects where physical circumstances present difficulties of adjoining proposals. Other situations, chiefly involving the recessed transverses and Park drives, have logical and defineable zonal divisions.

It is the objective of this proposal to dissect the Park in order to determine manageable projects of rehabilitation in a reasonable order of execution. The plan is open to modification and restructuring where dictated by the development of designs and progress in construction. The zonal conception to the program of rehabilitation has also assisted in the formulation of cost estimates and in determining a feasible scheduling of the quality, quantity and duration of the work.

# 12

# FISCAL CONCERNS

**George Rainer**

Flack & Kurtz Consulting Engineers

## FUNDING NEEDS

The development of new infrastructure is highly capital intensive. Program budgets are prepared at all levels of government for the construction of new roads, sewage treatment plants, transportation systems, and so on, and in due course, these facilities are generally built. However, the cost of maintaining these facilities is never included in the initial budgets; maintenance is always an operating expense that has to be borne by the local jurisdiction using the facility. During the life of the facility, which may vary from 30 to 60 years, there are always periods of fiscal stringency when savings measures must be instituted and funds earmarked for maintenance are taken out of the operating budget; that is, maintenance has been deferred. Once out of the budget, these funds are not restored and the facility suffers deterioration.

There is no clear-cut agreement on the magnitude of infrastructure repair and building needs in the United States; there have been a series of needs assessments during the early 1980s; these were conducted by a variety of organizations, including the Associated General Contractors of America, the Joint Economic Committee of Congress, and the Congressional Budget Office (CBO), among many others. They did not all consider the same infrastructure components, and they did not encompass the same time frames; therefore, the estimates range from $40 to $150 billion per year. Table 12-1 represent the CBO's 1983 estimate of annual spending needs and the federal share of the total. This summary shows that states and local governments will have to provide nearly 50% of the funds if the needs are to be met. In constant 1972 dollars, total U.S. infrastructure investment declined between 1960 and 1982 both in absolute terms ($22.8 billion to $19.9 billion) and on a per capita basis (from $126.20 to $85.50). During these same years, however, the federal share of these funds rose from 14.4% to 15.8% (Porter and Peiser, 1984).

**TABLE 12-1   ESTIMATED ANNUAL CAPITAL NEEDS FOR SELECTED INFRASTRUCTURE PROGRAMS, 1983–1990 (billions of 1982 dollars)**

| Infrastructure System | New Construction | Repair and Rehabilitation | Total | Federal Share |
|---|---|---|---|---|
| Highways | 9.9 | 17.3 | 27.2 | 13.1 |
| Public transit | 2.2 | 3.3 | 5.5 | 4.1 |
| Wastewater treatment | 6.1 | 0.5 | 6.6 | 4.2 |
| Water resources | 2.3 | 1.8 | 4.1 | 3.7 |
| Air traffic control | 0.1 | 0.7 | 0.8 | 0.8 |
| Airports | 1.0 | 0.5 | 1.5 | 0.9 |
| Municipal water supply | 3.6 | 4.1 | 7.7 | 1.4 |
| | 25.3 | 28.2 | 53.4 | 28.2 |

# SOURCES OF FUNDING

Most infrastructure facilities are used primarily at the local level and therefore should ideally be financed out of municipal funds. Frequently, municipal tax bases are too small to undertake major capital expenditures; and since local governments are creatures of the states with their much larger tax bases, financial assistance is generally sought at that level. Many public works issues spill over the borders of local jurisdiction and therefore take on a regional concern: sewage treatment, solid waste disposal, transit, and highways, for example. The state then becomes a much more effective arena for planning, budgeting, and funding these facilities.

The federal government's role is frequently one of setting standards: levels of sewage treatment to be attained, hazardous waste to be controlled, or drinking water quality to be maintained. However, when major policies are to be implemented in a certain time frame, major funding is provided at the federal level, as in the case of the interstate highway program or with the sewage treatment plant grant program. Certain infrastructure elements, such as communications, are always funded by the private sector; however, the practice of "privatization" of public works functions is being encouraged to reduce the necessity for public outlays. Table 12-2 gives a matrix of possible funding sources for various infrastructure elements.

There are approximately 3000 governmental entities that require funding for

## TABLE 12-2 POSSIBLE SOURCES OF FUNDING

| Infrastructure System | Federal | State | Local | Private |
|---|---|---|---|---|
| Water supply | × | × | × | × |
| Sewage treatment | × | × | × | × |
| Storm drainage | — | — | × | — |
| Solid waste | × | × | × | × |
| Power plants | × | × | × | × |
| Thermal plants | — | — | — | × |
| Alternative energy | × | × | — | × |
| Communications | — | — | — | × |
| Transit | × | × | × | × |
| Air traffic | × | — | — | × |
| Highways and bridges | × | × | × | — |
| Local roads | — | × | × | — |
| Waterfront | — | — | × | × |
| Ports and terminals | — | × | × | × |
| Waterways | × | × | — | — |

infrastructure projects. At any one time there can be as many as 400 federal categorical grant programs available to facilitate capital expenditures. These are administered by the Corps of Engineers, the Energy Department, and the Departments of Interior, Transportation, and Commerce (Hirten et al., 1983). Clearly, some coordination and planning are in order.

## ALLOCATION OF FUNDS

Allocation of funds occurs through policy, politics, or current vogue—usually, a combination of all three. When environmental protection was the watchword of the day during the 1960s, funds became available for sewage treatment plants. When energy was scarce during the 1970s, funds for research and development on alternative energy sources became abundant. The 1980s brought the realization that the nation's infrastructure was deteriorating; however, budget constraints mitigated against the forming of a federal infrastructure bank, which would have provided loans on a revolving basis. The federal gasoline tax was increased instead, and these funds have been beneficial in repairing highways throughout the United States.

## METHODS OF FUNDING

The basic method of funding public works is the issuing of bonds. General obligation bonds can be issued for any governmental purpose; interest and principal repayment are made out of general tax revenues. Revenue bonds are used when the purpose for which they are floated creates an income stream, or revenue, which is used to repay them over a period of years; funds for major toll roads are usually raised in this manner. However, revenue bonds are considered to carry higher risks than general obligation bonds and therefore carry higher interest rates. In recent years tax-exempt municipal bonds have become an important vehicle for financing infrastructure; these bonds carry lower interest rates than other obligations, but they appeal to high-income investors because the reduction in their tax obligation more than compensates for the lower interest earned. Local real property taxes go into a general fund, which is normally used to pay for current operations and for maintenance of existing infrastructure. There is no reason why these funds cannot be expended on capital projects—and they frequently are—but because of political considerations, alternative mechanisms have been developed.

Assessment districts, or public improvement districts, are areas in which it is permitted to levy special taxes on property owners because they benefit from a specific public improvement within the district, such as a parking garage that benefits a mixed-use development, or a pedestrian mall that enhances the business of abutting merchants. Even basic infrastructure such as a sewage treatment plants can be financed in this way if it used for overcoming a building moratorium in a given area, for example. These districts can be initiated by groups of property owners or by a public body, but a majority of the property owners affected must approve its creation. Local public officials usually favor this mechanism because it provides facilities at no direct cost to local government (Porter and Peiser, 1984).

Special districts operate somewhat differently, in that they are limited-purpose local governments that have the authority to tax, issue bonds, and provide services within a specified area. These governing bodies are considered "dependent" when they are controlled by a city or county, and "independent" when their bond issues are exempt from local statutory debt limits. They are known as *nonenterprise districts* if they levy taxes, and as *enterprise districts* if they collect user fees. The creation of special districts has been credited with defusing local political opposition to growth; the fact that the number of special districts has increased dramatically during the last decades attests to their popularity. Authorities set up to deal with specific functions, such as traffic, port management, or public development and funded off-budget, are examples of this type of arrangement.

In the case of tax-increment financing, the local government issues bonds to provide the infrastructure in a growth area, and then earmarks the increased taxes (above a set baseline) resulting from the new development to pay off the debt. In this way the new development pays for the support systems it requires. There may be nonphysical services, such as police and fire departments, which experience growing demand from the new development and do not benefit from the bond issue; but generally the mechanism is favored by developers and governments alike (Apogee Research, Inc., 1987).

As the federal government has attempted to reduce its grant programs to localities, local real estate taxes have had to pick up the shortfall. Resistance to these property taxes has fostered a movement toward user fees, whereby the beneficiaries of a given service pay for it directly. This includes water supply charges, boat docking fees, and park use fees, for example. Public officials consider the levying of user fees as being very efficient from the point of view of administration; but when it is considered that taxpayers of all income levels pay the same fee for basic services (such as garbage collection), it could be thought of as regressive taxation. In practice, however, user fees facilitate capital expenditures outside the framework of taxes and spending limits.

Development exactions have been levied for many years. It is common practice for developers in outlying districts to build roads, sewers, and drainage systems and then dedicate them to the locality. In recent times these exactions have included a broader range of facilities: A certain portion of the land being developed must be dedicated as a park, contributions to school building costs are required, or the modernization of adjacent subway stations is mandated as the price for being allowed to build at all. Although the community as a whole benefits from these mandated improvements, it is the buyers of the new homes being built, for example, who pay for the new infrastructure through the higher cost of their houses. It is a painless way to finance these necessary public improvements, however.

Impact or development fees are another method for getting the developer to share in the burden of public works costs. Charged at the time of issuing a building permit, these fees are meant to compensate the community for the additional burden on their public facilities caused by the development, such as intensified use of fire departments, police, or libraries. The advantages and disadvantages of impact fees are similar to those of exactions. They have been used with some success in local growth management programs by differentiating among the fees for specific types of development, thus favoring some and discouraging others (Apogee Research, Inc., 1987).

### Federal and State Assistance

Throughout this book there have been references to federal and state assistance to localities. This has come in the form of loans or outright grants and has become a major factor in preparing local budgets. There has always been an attempt to target this assistance to specific needs, both social and physical. During the 1970s the federal government decided to share its immense tax base with the localities directly; revenue sharing was disbursed to the localities directly in proportion to their population and funds could be used for almost any purpose. The 1980s brought Urban Development Action Grants (UDAG) targeted to areas requiring economic development. The rules for participation were specific, and a clear benefit to disadvantaged neighborhoods was mandatory.

During economic downturns the development of public works provides several benefits; it employs many construction workers and provides infrastructure to areas where private development could not occur otherwise. The late 1970s saw massive infusions of federal funds into such stimulative public works projects (Choate and Walter, 1983). Although these assistance programs are desirable in and of themselves, they are subject to the political outlook of the moment and do not form a firm basis for infrastructure planning.

Some states have considered the institution of infrastructure banks that would set up revolving funds to provide seed money for necessary projects. Localities would borrow funds at no interest and repay the loans out of user charges. The source of initial funding would be a combination of bond issues, federal revenue sharing, and possibly private capital.

## Privatization

Private ownership of public facilities is becoming a possibility that is attracting increasing attention. Private owners can take tax credits for depreciation of facilities, which provides tax advantages, even though their borrowing costs would be higher. The locality obtains the facility without capital outlays and pays only an annual service or maintenance fee. This type of arrangement works better with some public facilities than with others: Garbage collection is frequently handled by private contractors; bus fleets can be leased by the year; and sports facilities are financed by private capital. However, there is less interest by private owners to build highways or operate sewage treatment plants; but if the need exists and tax laws remain favorable to the method, there is no limit to the possibilities.

There is an opportunity for combined financing between private and public entities. Funding and benefits can be shared, as in the case of a lease-back arrangement, or initial private ownership can revert to the public agency after a period of years. The primary interest is to minimize capital expenditure (and borrowing) while obtaining the use of the required infrastructure in a cost-effective manner.

# ECONOMIC DEVELOPMENT

The infrastructure that was built in the early years of this country's development—the canals, the dams, the railroads—has supported the phenomenal economic growth that has taken place in the United States. By allowing existing facilities to deteriorate, and by underinvesting in new ones, we are seriously threatening our future economic development. Quite apart from the health and safety issues involved, the economic feasibility of a region is greatly influenced by the condition of its infrastructure. Roads and bridges, for example, are needed by businesses and the general public alike; where they are lacking, goods deliveries are handicapped. Where maintenance has been neglected, vehicle repair costs to users have increased substantially (Associated General Contractors of America, 1984). New factories will not locate in an area where there is no water and sewer service; when existing water mains fail, workers are out of jobs. These lifelines are closely tied to our economy.

The level of public spending is a clear indicator of the quality of the services we receive. Since 1960, government capital investment has dropped from 3.2% of GNP to 1.3% in 1982. It is known that public investment generally increases the productivity of private investment; the growing imbalance between the two sectors may mean that we are not benefitting from the private dollar investment to the extent that we should, which undoubtedly impairs our competitive edge in world trade. Public works construction has often been used as a counter-cyclical measure during recessions by providing many jobs; but even during normal economic times, contract construction in public works can contribute as many as 50% of all jobs in the construction trades (Associated General Contractors of America, 1984). It is imperative that our infrastructure be rebuilt if the economy is to remain healthy.

## LEGISLATION

In 1984 the Public Works Improvement Act (P.L. 98-501) created a National Council on Public Works Improvement which undertook an assessment of the U.S. infrastructure. A nine-volume report was completed in 1987; in its final report in 1988 it gave a report card on all types of facilities and made recommendations for the future. It then went out of business. The report card looked as follows:

| | | | |
|---|---|---|---|
| Highways | C+ | Water supply | B− |
| Mass transit | C− | Wastewater | C |
| Aviation | B− | Solid waste | C− |
| Water resources | B | Hazardous waste | D |

This clearly shows where our emphasis on spending should be placed, and it will be up to the federal government to set the policy in this regard. Out of 38 infrastructure-related bills that were introduced in Congress in 1983, four bills with minimal funding were passed (AGC 1984). Clearly, the impetus will have to come from states and localities if any progress is to be made. None of the three broad sources of local and state infrastructure funding—current state and local government revenues, debt issuance, and federal grants—are likely to increase in the immediate future. Unless the political and economic climate reverses these trends, the future for improvements in the present condition of the nation's infrastructure looks very poor indeed.

## REFERENCES

Apogee Research, Inc., *Financing Infrastructure: Innovation at the Local Level*, National League of Cities, Washington, D.C., 1987.

Associated General Contractors of America, *Infrastructure: The Aftermath and the Realization*, AGCA, Washington, D.C., 1984.

Choate, Pat, and Susan Walter, *America in Ruins: The Decaying Infrastructure*, Duke University Press, Durham, N.C., 1983.

Hirten, John, et al., eds., *An Infrastructure Planning Process for the U.S.*, K.C. 1192, Department of HUD, Washington, D.C., 1983.

Porter, Douglas R., and Richard B. Peiser, *Financing Infrastructure to Support Community Growth;* ULI Development Component Series, Urban Land Institute, Washington, D.C., 1984.

## ADDITIONAL READING

Alterman, Rachelle, ed., *Private Supply of Public Services: Evaluation of Real Estate Exactions*, New York University Press, New York, 1988.

Congressional Budget Office, *Public Works Infrastructure: Policy Considerations for the 1980's*, U.S. Government Printing Office, Washington, D.C., 1983.

DeChiara, Joseph, and Lee Koppelman, *Urban Planning and Design Criteria*, 3rd ed., Van Nostrand Reinhold, New York, 1982.

Frank, James, and Robert M. Rhodes, eds., *Development Exactions*, APA Planners Press, Chicago, 1987.

Hatry, H. P., and G. E. Peterson, *Maintaining Capital Facilities: Executive Report*, Urban Institute, Washington, D.C., 1983.

Kraft, and Brown, *Rebuilding America: Infrastructure Rehabilitation*, American Society of Civil Engineers, New York, August 1984.

National Council on Public Works Improvement, *Fragile Foundations: A Report on America's Public Works*, Final Report, NCPWI, Washington, D.C. February, 1988.

## QUESTIONS

1. According to the Congressional Budget Office, what is the total estimated funding need for infrastructure?

2. What proportion will have to be funded out of state and local funds?

3. Which element of infrastructure is always funded by the private sector? Which never?

4. How are infrastructure funds allocated?

5. Describe two types of bond issue.

6. Discuss at least two types of off-budget sources of funding, and name the advantages of each.

7. Describe two types of federal funding methods widely used during the 1970s and 1980s.

8. What are the advantages and disadvantages of privatization?

9. How does infrastructure funding affect economic development?

10. What is the infrastructure element requiring the greatest attention? How did you arrive at this conclusion?

# INDEX

combustion of, as energy source, 51–52
costs of, 40
disposal of, 37–38
in parks, 246–247
planning considerations, 38–39
quantities generated, 33–34
treatment of, 36–37
types of, 32–33
Solid Waste Disposal Act of 1965, 42
Special Bridge Replacement Program,
140–141
Sprint, 79
Starved air incineration, of solid waste, 52
Storm drainage, 16
design of, 17–19
water quality of, 20
Streets:
and buses, 229
classification of, 103–104
design of:
drainage, 108–109
intersections, 107
lighting, 110
right-of-way, 106–107
surveys, 105
traffic analysis, 105–106
traffic signs, 110–111
width, 107
downtown plans, 104–105
history of, 100–102
landscaping of, 111–113
planning considerations:
bicycle paths, 119
downtown plan, 113–118
highway integration, 119
life cycle, 113
pedestrians, 119
urban design guidelines, 119–123
purpose of, 102
refuse from, 32
space allocation for, 103
use of, 102
Subways, *see* Transit systems
Superfund Program, 42
Surcharging, of sewers, 16
Surface Transportation Assistance Act of 1978,
140

Tacoma Narrow Bridge, 142–143
Telecommunications, 76
bypass technology in, 82

cable plant in, 81–82
cellular telephones in, 8
central office role in, 79–80
digital service, 86–88
future of, 92–95
history of, 77–79
integrated services digital network, 89–90
interconnect companies in, 78
microwave systems in, 82–83
planning considerations in, 88–89
private branch exchange in, 80–81
satellite systems in, 83–85
service types in, 86–88
standards for, 92
switching systems in, 77, 79–81
technology of, 79–88, 89–91
very small aperture terminal, 91
video teleconferencing, 91
voice service, 86
Teleport, 85
Terminal design, airport, 212–213
Thermal energy, 51–52, 56
Title III, in hazardous waste management, 42
Ton (cooling unit), 58
T-1 signal, 87
Topographic survey, in street design, 105
Traffic signals, 110–111
Traffic survey, in street design, 105–106
Transfer stations, 35
Transit systems:
classification of, 191–192
costs of, 198–200
and energy use, 65–71
modal choices of, 192–193
peaking problems in, 195–196
performance criteria, 193–194
population density and, 194–195
privatization of, 200–201
role of, 190–191
system components, 196–198
urban design issues in, 201–203
Transponders, 84
Transportation Research Board, 206

U.S. Public Health Service, 12
Urban areas, highways in, 119
Urban design guidelines, 119–123
Urban Land Institute, 96
Urban Mass Transportation Administration,
199, 200
Utility survey, in street design, 105